Yeddanapudi Swamy

Solid and Hazardous Waste Management Practices On Board Cargo Vessels

Yeddanapudi Swamy

Solid and Hazardous Waste Management Practices On Board Cargo Vessels

LAP LAMBERT Academic Publishing

Impressum / Imprint

Bibliografische Information der Deutschen Nationalbibliothek: Die Deutsche Nationalbibliothek verzeichnet diese Publikation in der Deutschen Nationalbibliografie; detaillierte bibliografische Daten sind im Internet über http://dnb.d-nb.de abrufbar.
Alle in diesem Buch genannten Marken und Produktnamen unterliegen warenzeichen-, marken- oder patentrechtlichem Schutz bzw. sind Warenzeichen oder eingetragene Warenzeichen der jeweiligen Inhaber. Die Wiedergabe von Marken, Produktnamen, Gebrauchsnamen, Handelsnamen, Warenbezeichnungen u.s.w. in diesem Werk berechtigt auch ohne besondere Kennzeichnung nicht zu der Annahme, dass solche Namen im Sinne der Warenzeichen- und Markenschutzgesetzgebung als frei zu betrachten wären und daher von jedermann benutzt werden dürften.

Bibliographic information published by the Deutsche Nationalbibliothek: The Deutsche Nationalbibliothek lists this publication in the Deutsche Nationalbibliografie; detailed bibliographic data are available in the Internet at http://dnb.d-nb.de.
Any brand names and product names mentioned in this book are subject to trademark, brand or patent protection and are trademarks or registered trademarks of their respective holders. The use of brand names, product names, common names, trade names, product descriptions etc. even without a particular marking in this works is in no way to be construed to mean that such names may be regarded as unrestricted in respect of trademark and brand protection legislation and could thus be used by anyone.

Coverbild / Cover image: www.ingimage.com

Verlag / Publisher:
LAP LAMBERT Academic Publishing
ist ein Imprint der / is a trademark of
AV Akademikerverlag GmbH & Co. KG
Heinrich-Böcking-Str. 6-8, 66121 Saarbrücken, Deutschland / Germany
Email: info@lap-publishing.com

Herstellung: siehe letzte Seite /
Printed at: see last page
ISBN: 978-3-659-39827-8

Solid and Hazardous Waste Management Practices On Board Cargo Vessels

Yeddanapudi VRPP Swamy

From the Chief Engineer, M.V. TAKARA,
Wilhelmsen Lines Car Carriers Ltd. (WLCC),
Southampton,
United Kingdom.

This book is for my wife,

YASMIN

"the guiding light"

Contents

Abstract

Shipping or carriage of goods plays an important role in the development of human societies and international shipping industry, which carries 90% of the world trade, is the life blood of global economy. During ships operational activity a number of solid and hazardous wastes, also referred as garbage are produced from galleys, crew cabins and engine/deck department stores, etc. The present review provides an overview of the current practices on board a cargo vessel, and examines the evidence that links the waste management plan regulations to shipping trade. With strict compliance to International Maritime Organization's 'International Convention for the Prevention of Pollution from Ships', called MARPOL Convention, which prevents the pollution of sea from ships various discharges, a well documented solid and hazardous waste management practices are being followed on board ships, in accordance with Annex V of the convention. With the most recently revised MARPOL Annex V, enforced from 1st January 2013, new requirements have also been added prohibiting almost all discharges of waste into the sea. All ship board wastes are collected, segregated, stored and disposed properly in appropriate locations, as per on board solid and hazardous waste management plan, in accordance with the shipping company's environmental protection policy. For instance, food residues are ground on board and dropped into the sea as fish food. Cardboard and the like are burned on board in incinerators. Glass is sorted into dark/light colours and deposited ashore, as are plastics, metal, tins, batteries, fluorescent tubes, etc. The residue from plastic incineration which is still considered as plastic is brought back for shore disposal to port reception facilities. On the other hand, recyclable material identified is stored separately in green coloured receptacles for shore disposal. New targets are being set up to reduce the volume of garbage generated as well as disposed to shore facilities, and new ships are using baling machines which compress cardboard etc. into bales to be taken ashore. However, this solid and hazardous waste management and its control system work on a 'continual improvement' process for a sustainable future.

Key words: Shipping, solid and hazardous waste management plan, garbage, regulations, storage, disposal practices, en route, documentation, implementation.

1. Introduction

Today, the international shipping industry carries 90% of the world trade and is the life blood of global economy.[1] There are around 79,074[2] merchant ships (Table 1), flying the flags of over 150 different nations trading internationally and transporting every type of cargo. Without shipping, inter-continental trade, both transporting raw material and import and export of affordable food, and manufactured goods is simply not possible. Sea borne trade thus continues to expand, bringing benefits to the customers across the world through competitive freight costs. With the increased economic liberalisation and the growing efficiency of shipping as a mode of economic transport, prospects of the international shipping industry's further growth remains strong.[3]

During ships operational activity a number of solid and hazardous wastes also referred as garbage, are produced from galleys, crew cabins and engine/deck department stores, etc. Table 2 describes the volume of garbage generated in a year on a typical ocean going car carrier ship. Indeed, the amount of garbage depends on persons on board a ship, length of each voyage, the on board application of minimisation technology, type of ship and the ships operational conditions. Whatever waste is generated like plastic, oily rags, oily waste, used batteries, used florescent tubes and bulbs, and incinerated ash of any plastic products is considered as hazardous waste and these waste should not be disposed in to sea at any time.

It is interesting and shocking to learn that a plastic bottle, if thrown into the sea takes 450 years[4] to dissolve in the sea as illustrated in Table 3. Therefore, to protect our seas, solid and hazardous waste generated should be collected, segregated, stored and disposed properly, in accordance with the shipping company's environmental protection policy and as per the International standard maritime (ISM)[5] code of the shipping company. In the present review, current systems of solid and hazardous waste management plan on board will be discussed, and the potential implementation procedures and controlling measures are highlighted. The current evidence is presented in three main sections: first the solid and waste management practices on board are described, followed by the various implementation and/or controlling procedures that regulate the management plan, and lastly some proposals to the environmental management system in way of reducing amount of waste generated and future targets are discussed. Tables 4-16 and Figures 1-4 summarize all the current findings that support the 'solid and hazardous waste management plan' on board a cargo ship.

1.1 *Solid and Hazardous Waste Management Practices On Board Ocean Going Ships*

Solid and hazardous waste management practices include sorting of the garbage, by way of collection and separation; garbage reduction by using the on board processing units and equipment,

1

and storage for disposal as per the environmental policy of the company. Motivating educating and imparting continuous training, to the crew, in meeting the regulatory requirements is integral part of on board solid and hazardous waste management practise. Moreover, health, safety of the crew and the protection of the environment are given priority while developing procedures and practises in waste management plans.

1.1.1. Solid and Hazardous Waste Generated On Board Ship and Their Definitions Along with Other Special Definitions Relating to Waste Management Practises

In general, on board any ocean going merchant ship, a number of solid and hazardous wastes also referred as garbage, are generated and these are defined below:

Hazardous waste means, any waste which, due to its nature, physical, chemical, or infectious properties, is potentially hazardous to human health and /or the environment during its use, handling, storage or transportation. It includes any material which may require special handling, disposal or recycling techniques to eliminate or reduce the hazard.

Harmful substance means, any substance which, if introduced into the sea is liable to, in creating hazards to human health, harming living resources and marine life, damaging amenities or to interfering with other legitimate uses of the sea.

Effluent means, discharged liquid which may contain harmful substances /residues in solution or in suspension.

Discharge means, in relation to harmful substances or effluents, containing such substances, any release, howsoever caused, from a ship including any escape, disposal, spilling, leaking, pumping, emitting or emptying.

Garbage means all kinds of food wastes, domestic wastes and operational wastes. All plastics, cargo residues, incinerator ashes, cooking oil, fishing gear, and animal carcasses generated during the normal operation of the ship and liable to be disposed of continuously or periodically. However, garbage does not include fresh fish and parts thereof generated as a result of fishing activities undertaken during the voyage, or as a result of aquaculture activities involved with the transport of fish including shellfish for placement in the aquaculture facility and transport of harvested fish including shellfish from such facilities to shore for processing.

Animal carcasses means the bodies of any animals that are carried on board as cargo and that die or are euthanized during the voyage.

Cargo residues means the remnants of any cargo which remain on the deck or in holds, following loading or unloading, including loading and unloading excess or spillage, whether in wet or dry condition or entrained in wash water but does not include cargo dust remaining on the deck after sweeping or dust on the external surfaces of the ship.

2

Cooking oil means any type of edible oil or animal fat used or intended to be used for the preparation or cooking of food, but does not include the food itself that is prepared using these oils.

Dishwater means the residue from the manual or automatic washing of dishes and cooking utensils which have been pre-cleaned to the extent that any food particles adhering to them would not normally interfere with the operation of automatic dishwashers.

Grey water means drainage from dishwater, shower, laundry bath and washbasin drains. It does not include drainage from toilets, urinals, hospitals, and animal spaces, and it does not include drainage from cargo spaces.

Domestic waste means all types of wastes that are generated in the accommodation spaces on board the ship. Domestic waste does not include grey water.

Food waste means any spoilt or unspoilt food substances and includes fruits, vegetables, dairy products, poultry, meat products and food scraps generated on board ship.

Incinerator ash means ash and clinkers resulting from shipboard incinerators used for the incineration of garbage.

Maintenance waste means materials collected in the engine and deck departments while maintaining and operating the vessel, such as soot, machinery deposits, scraped paint, deck sweeping, wiping wastes, and rags, etc.

Non-recyclable plastic means any type of plastic that has NO number for recycling (eg. trash bags, zip lock bags, inside cereal box plastic, bubble wrap, clear plastic wrap, soiled plastic bottles and bags etc.).

Oily rags are rags which have been saturated with oil.

Contaminated rags are rags which have been saturated with a substance defined as a harmful substance.

Operational waste means all solid wastes (including slurries) that are collected on board during normal maintenance or operations of a ship, or used for cargo stowage and handling. Operational waste also includes cleaning agents and additives contained in cargo hold and external wash water. Operational wastes does not include grey water, bilge water, or other similar discharges essential to the operation of a ship, taking into account the guidelines developed by the organization.

Plastic means a solid material which contains as an essential ingredient one or more high molecular mass polymers and which is formed (shaped) during either manufacture of the polymer or the fabrication into a finished product by heat and/or pressure. Plastics have material properties ranging from hard and brittle to soft and elastic. For the purposes of this regulation, 'plastics' mean all garbage that consists of or includes plastic in any form, including synthetic ropes, synthetic fishing nets, plastic garbage bags and incinerator ashes from plastic products.

3

Medical waste means any solid waste that is generated in the diagnosis, treatment or immunization of human being or animals, in research pertaining thereto, or in the production or testing of biological, including bit not limited to isolation wastes, infectious agents, human blood and blood products, pathological wastes, sharps, body parts, contaminated bedding, surgical wastes, potentially contaminated laboratory and dialysis wastes. Generally medical waste is distinguished into two categories: infectious and non-infectious.

Quarantine waste means any solid or liquid waste determined by local or regional legislation to require special handling, segregation and disposal due to its potential to spread disease, diseases or plant and animal pest when discharged.

Recyclable material means a material or product that is able to be recycled.

Recycling means the activity of segregating and recovering components and materials for reprocessing.

Reuse means the activity of recovering components and materials for further use without reprocessing.

Other special definitions for the purpose of waste management practises are:

En route means that the ship is underway at sea on a course or courses, including deviation from the shortest direct route, which as far as practicable for navigational purposes, will cause any discharge to be spread over as great an area of the sea as is reasonable and practicable.

Fishing gear means any physical device or part thereof or combination of items that may be placed on or in the water or on the sea-bed with the intended purpose of capturing, or controlling for subsequent capture or harvesting, marine or fresh water organisms.

Fixed or floating platforms means fixed or floating structures located at sea which are engaged in the exploration, exploitation or associated offshore processing of sea-bed mineral resources

Nearest land means from the baseline from which the territorial sea of the territory in question is established in accordance with international law, except for the purposes of the present Annex. For example, 'from the nearest land' off the north eastern coast of Australia shall mean from a line drawn from a point on the coast of Australia in:

Latitude 11°00′S, longitude 142°08′E

to a point in latitude 10°35′ S, longitude 141°55′ E,

thence to a point latitude 10°00′ S, longitude 142°00′ E,

thence to a point latitude 09°10′ S, longitude 143°52′ E,

thence to a point latitude 09°00′ S, longitude 144°30′ E,

thence to a point latitude 10°41′ S, longitude 145°00′ E,

thence to a point latitude 13°00′ S, longitude 145°00′ E,

thence to a point latitude 15°00′ S, longitude 146°00′ E,

thence to a point latitude 17°30' S, longitude 147°00' E,

thence to a point latitude 21°00' S, longitude 152°55' E,

thence to a point latitude 24°30' S, longitude 154°00' E,

thence to a point on the coast of Australia in latitude 24°42' S, longitude153°15' E

Special area means a sea area where for recognized technical reasons in relation to its oceanographic and ecological conditions and to the particular character of its traffic the adoption of special mandatory methods for the prevention of sea pollution by garbage is required.

For the purposes of this Annex the special areas are the Mediterranean Sea, the Baltic Sea, the Black Sea, the Red Sea, the Gulfs, the North Sea, the Antarctic and the Wider Caribbean regions which are defined as follows:

The Mediterranean Sea area means the Mediterranean sea proper including the gulfs and seas therein with boundary between the Mediterranean and the Black sea constituted by the 41° N parallel and bounded to the west by the Straits of Gibraltar at the meridian 5°36' W

The Baltic Sea area means the Baltic Sea proper with the Gulf of Bothnia and the Gulf of Finland and the entrance to the Baltic Sea bounded by the parallel of the Skaw in the Skagerrak at 57° 44.8' N.

The Black Sea area means the Black sea proper with the boundary between the Mediterranean and the Black seas constituted by the parallel 41° N.

The Red Sea area means the Red sea proper including the Gulfs of Suez and Aqaba bounded at the south by the rhumb line between Ras si Ane (12° 28.5' N, 43° 19.6' E) and Husn Murad (12° 40.4' N, 43° 30.2' E).

The Gulfs area means the sea area located north-west of the rhumb line between Ras al Hadd (22° 30' N, 59° 48' E) and Ras al Fasteh (25° 04' N, 61° 25' E).

The North Sea area means the North sea proper including seas therein with the boundary between:

the North sea southwards of latitude 62° N and eastwards of longitude 4° W;

the Skagerrak, the southern limit of which is determined east of the Skaw by latitude 57° 44.8' N; and

the English Channel and its approaches eastwards of longitude 5° W and northwards of latitude 48° 30' N.

The Antarctic area means the sea area south of latitude 60° S.

The Wider Caribbean Region means the Gulf of Mexico and Caribbean sea proper including the bays and seas therein and that portion of the Atlantic Ocean within the boundary constituted by the 30° N parallel from Florida eastward to 77°30' W meridian, thence a rhumb line to the intersection of 20° N parallel and 59° W meridian,

thence a rhumb line to the intersection of 7°20' N parallel and 50° W meridian,

thence a rhumb line drawn southwesterly to the eastern boundary of French Guiana.

Port Reception Facilities mean fixed, floating or mobile port facility for the reception of ship generated residues /wastes. This relate to reception facilities for garbage as defined by revised MARPOL Annex V.

Environmental Aspects mean elements of an organization's activities, products or services that can interact with the environment.

Environmental Impacts mean any changes to the environment whether adverse or beneficial wholly or partly resulting from an organization's activities, products or services.

1.1.2   Solid and Hazardous Waste Categories

For the sake of standardization and to keep/record correct documentation of different wastes generated in ocean going ships, the solid and hazardous waste generated on board is categorised into nine groups for the entire shipping industry. The waste categories are generally mentioned in the ship's solid and hazardous waste management plan and in the garbage record book (GRB) are given below.

Category A: Plastics and medical waste includes synthetic ropes, synthetic fishing nets, plastic garbage bags and incinerator ashes from plastic products which may contain toxin or heavy metal residues.

Category B: Food waste

Category C: Domestic wastes (e.g., paper products, rags, metal, boltless crockery etc.)

Category D: Cooking oil

Category E: Incinerator ashes

Category F: Operational wastages (floating dunnage or material, lining or packing material)

Category G: Cargo residues

Category H: Animal carcass

Category I: Fishing gear.

It is worth noting that before 1st Jan 2013, all the wastes were categorised as 1 to 6, where category 1 included plastic, 2 floating dunnage, lining or packing material, 3 ground paper products, rags, glass, metal, bottles, crockery etc., 4 cargo residues paper products, rags, glass, metal, bottles, crockery etc., 5 food waste, and 6 as incinerated ash. But after 1st Jan 2013, when the new revised MARPOL Annex V came into force, the categorization was different, with the inclusion of extra groups as described in the above paragraph with specific descriptions for each of these categories (A to I). Furthermore, an 'en route' clause was also introduced which meant that most allowable discharges like comminuted or ground food waste, are only permitted while the ship is in motion. This implies stricter compliance from the global shipping, in order to reduce the impacts of their operational discharges on seas and beaches.

2   Objective of Solid and Hazardous Waste Management Plan

The functional activities of the solid and hazardous waste management and its control system are illustrated in Figure 1. Like any management control system, this solid and hazardous waste management system on board a ship has a plan, with on board designated organizations working under the regulations of the revised MARPOL Annex V,[6] and the objective is controlled by inspections and audits on a continuous basis. This revised Annex V deals with different types of garbage and specifies the distance from land and the manner in which they should be disposed. The requirements are much stricter in a number of special areas, but perhaps the most important features of this revised Annex is complete ban imposed on dumping of all forms of plastic into the seas even in incinerated ash form.

The aim of this plan is to provide ship's staff with best waste management practices in order to comply with the company's environmental protection policy and regulatory requirements of revised MARPOL Annex V. It explains about the placards, garbage management plans and garbage record keeping mentioned in all categories of garbage generated on board. Compliance with the provisions of revised Annex V will require careful planning and proper execution by shipboard personnel. This plan details the most appropriate procedures for collecting, storing, processing and disposing ship generated garbage, including the appreciation of the use and limitation of processing equipment fitted on board. Every garbage management plan along with garbage record book will be inspected by a competent authority (port state or MARPOL inspectors) of the government which is a party to the convention.

3   Implementation of Solid and Hazardous Waste Management Plan

Protecting global environment is everybody's responsibility. Similarly, keeping oceans and beaches clean is the foremost duty of every crew member of the ship. This should be done by adopting proper implementation of the solid and hazardous waste management plan and the necessary strategies. Figure 2 shows a typical on board organization chart, to understand the duties and responsibilities of crew working on board. Master of the vessel has the sole responsibility for the safe and environmental friendly operation of the ship. The master is assisted by two departmental

7

heads, namely engine department head (chief engineer) and deck department head (chief officer). Under the chief engineer there are other engineering officers assisted by engine ratings and cadets for day to day operation and maintenance of on board machinery which includes compactors, incinerators etc. Similarly, under the chief officer there are deck officers who assist the chief officer in cargo operations and cargo related machinery maintenance, with the help of deck ratings and deck cadets. In fact, every person is a part of the safety, quality and environmental management system on board the ship.

During implementation of this waste management system four main steps are involved. They are collection and separation, processing, storing in the designated areas and disposing. Even while transporting these different categories of waste, identified and designated containers are used on board to avoid mixing of different categories of waste while transporting to the environmental station (storage location) in the ship.

3.1   *Collection and Separation of Solid and Hazardous Waste*

In general, respective heads of department of the vessel are responsible for the collection, separation and transportation of garbage generated from their areas of work to the designated storage area. In addition, every individual is solely responsible to segregate garbage at its source to prevent mixing during storage and disposal. The chief officer is specifically responsible for handling category E waste, incinerated ash which is hazardous. Because the residue from the plastic incineration (if the incinerator is approved by authorities to burn on board generated plastic) is still considered plastic and therefore it cannot be discharged overboard. It is only discharged to shore reception facilities with a proper receipt. The department head of each respective area ensures that suitable receptacles are provided and located to facilitate the efficient collection and separation of garbage in accordance with the garbage categories. The department head of each area supervises and trains their respective staff in the collection and separation as well as transportation of garbage in designated containers, to the designated area for further processing, storage or disposal as applicable. Table 4 shows the areas of responsibility for collection and separation of solid and hazardous waste on board an ocean going ship.

3.2   *Various Equipment Used and Processing of Solid and Hazardous Waste Generated*

Table 5 provides the description and probable locations of the equipment utilised in processing various garbage. The department heads are responsible for ensuring that the personnel operating and maintaining the equipment listed above are properly trained to do so. However, the chief engineer is consulted for detailed operational instructions and maintenance manuals. Wastes that are not

8

incinerated are stored until there are disposed to facilities ashore. Some functions and characteristics of the various processing devices used are as given below.

Shredders: A dry garbage shredder is used to increase the density of the garbage by shredding the waste components. The volume reduction depends on the material to be shredded. The garbage is commonly shredded to grain size elements. For storage on board the ship, shredded garbage may be compacted and the liquid extracted so as to avoid odour and hygiene problems.

Compactors: Compactors make garbage easier to store, to transfer to port reception facilities and to dispose of at sea when discharge limitations permit. The importance of compactors in waste management control system is given in Table 6.

Comminuter/grinders: Ships operating primarily beyond 3 nautical miles from the nearest land are encouraged to install and use comminuters to grind food wastes to a particle size capable of passing through a screen with openings no larger than 25 millimetres. Such a process is recommended even beyond 12 nautical miles because the particle size hastens assimilation into the marine environment.

Incinerators: Shipboard incineration is allowed only in a shipboard incinerator. Incineration of ship generated waste is prohibited in the Baltic Sea area under the Baltic Marine Environment Protection Commission.

Shipboard incineration of the following substances is prohibited:

- residues of cargoes subject to Annex I, II or III[7] or related contaminated packing materials
- polychlorinated biphenyls (PCBs)
- garbage, as defined by Annex V, containing more than traces of heavy metals
- refined petroleum products containing halogen compounds
- sewage sludge and sludge oil either of which are not generated on board the ship, and
- exhaust gas cleaning system residues.

Shipboard incineration of polyvinyl chlorides (PVCs) is prohibited, except in shipboard incinerator for which an IMOs type approval certificates are issued.

Shipboard incineration of sewage sludge and sludge oil generated during normal operation of a ship may also take place in the main or auxiliary power plant or boilers, but not inside ports, harbours and estuaries.

Each incinerator on a ship constructed on or after 1st January 2000 or incinerator which is installed on board a ship on or after 1st January 2000, should meet the requirements contained in MARPOL Annex VI appendix IV[8]. Each incinerator subjected to this subparagraph will be approved by the administration taking into account the standard specification for shipboard incinerators developed by the organization. The administration may allow exclusion to any incinerator which is installed on board a ship before 19th May 2005, provided that the ship is solely engaged in voyages within waters subject to the sovereignty or jurisdiction of the State flag,[9] which the ship is entitled to fly.

Incinerators installed in accordance with the requirements MARPOL Annex VI Appendix IV shall be provided with a manufacturer's operating manual which is to be retained with the unit and which shall specify how to operate incinerator within the approved limits. Moreover, personnel responsible for the operation of an incinerator shall be trained to implement the guidance provided in the manufacturer's operating manual.

The combustion chamber gas outlet temperature is monitored at all times the unit is in operation. Where that incinerator is of the continuous feed type, waste shall not be fed into the unit when the combustion chamber gas outlet temperature is below 850°C. But where that incinerator is of the batch loaded type, the unit shall be designed so that the combustion chamber gas outlet temperature shall reach 600°C within five minutes after start-up and will thereafter stabilize at a temperature not less than 850°C.

The IMO type approval certificate for shipboard incinerators with capacity of up to 1,500kW, should have the following statement: "This is to certify that the shipboard incinerator listed has been examined and tested in accordance with the requirements of the Standard for Shipboard Incinerators for disposing of ship-generated waste appended to the Guidelines for the Implementation of Annex V of MARPOL 73/78 as amended by resolution MEPC.76 (40) and referenced by regulation 16 of Annex VI to MARPOL 73/78."

Due to the potential environmental and health effects from combustion of by products eg. scraped paint, impregnated wood and PVC-based plastics, special precaution is required. Special rules on incineration may be established by authorities in some ports and may exist in some special areas. The incinerator shall not be operated in port to avoid air pollution hazards by smoke. Table 7 explains some built-in characteristics of the shipboard incinerators and procedures involved in their operation. The personnel who are in charge of the operation and maintenance of this equipment should be aware of these options for a safe and environmental friendly operation. In the shipboard International Oil Pollution Prevention Certificate, the incinerator capacity should be mentioned as KW or Kcal/hr heat units[10] as per the new requirements.

3.3   *Storage and Disposal of Garbage Solid and Hazardous Waste*

Table 8 provides an overview of management of all types of solid and hazardous waste generated on board an ocean going vessel. From this table, it can be seen that except for food waste, all solid and hazardous waste are to be brought back from sea to shore, and deposited with shore reception facilities with proper documentation.

However, for storage purposes, separate bins for disposal will be placed on board in key storage areas. The locations for garbage storage areas on board generally should meet the following requirements.

- Access to the site shall be free from obstructions, as far as practicable
- The transport route to manually land the garbage to shore shall be free from thresholds, coamings, and other obstructions, as far practicable
- A means for securing the storage containers against movement in the site shall be provided
- Storage sites, associated passageways, shafts and hatchways for vertical transport, and entrances shall be adequately sized for easy use, handling, and transport of storage containers
- Relevant fire protection equipment shall be provided at the storage sites
- For internal sites ventilation with a forced exhaust and natural supply with at least 5 complete air exchanges per hour shall be provided
- A water connection shall be provided for wet cleaning
- Inside scuppers shall be provided with a strainer. Wash water and escaping liquids from inside spaces shall be directed to an appropriate waste liquid system
- Separate storage sites or rooms should be considered for hazardous material storage. These spaces shall have drainage relevant to the material being stored and shall have an eyewash station for personnel in a readily accessible location.

The colour codes for the receptacles are used as an efficient means of maintaining segregated storage. Garbage is to be stored in marked/coloured bins segregated for disposal at sea or discharged to shore reception, described in Table 9. If receptacles such as drums are used, then they must be kept closed with a lid at all times. The garbage storage site should also be equipped with a,

- suitable absorbent material for oil-containing waste
- temporary storage in the event of broken containers e.g., pans and barrels
- broom, shovel
- locks, locking strips, cargo netting or other suitable protections against sliding, tilting, leaking or falling of stored garbage

- first aid kit
- sorting and handling procedures
- machinery operating instructions, and
- adequate lighting

If the storage area is located on weather deck then,

- storage sites should be sheltered from weather and seawater as far as possible
- deck storage site shall be permanently marked and be of sufficient size to accommodate the garbage containers
- location of the garbage storage site(s) shall be appropriately selected according to categories of garbage to be stored, and located so as not to interfere with normal vessel operations, and
- means for securing outside garbage containers against movement shall be provided.

More importantly, the selection of storage area to be documented in the solid and hazardous waste management plan on board and all crew should be aware of the location and different coloured receptacles. The receptacles should be properly covered and with proper lids on top of them. Quarantine officers are very particular about this aspect and strict inspections will be carried out. In USA, if the lids are not properly closed or any insects are found in the receptacle drums, then quarantine officials will levy heavy fines to the vessels and some times even stop the cargo loading / discharging activities till the required corrective action is taken by the vessel management to the satisfaction of the inspecting authorities. The next paragraph explains the violation and correct procedure required to avoid any recurrences.

3.3.1 Deficiencies Noted by Authorities As Far As Violation of MARPOL Annex V

Place: USA Port Newark; Date 25th Oct 10; Vessel Type - PCC; 2 Deficiencies, vessel fined USD1000

Deficiency 1: One empty regulated garbage drum in upside down position on open deck.
Deficiency 2: One wooden box on open deck not sealed with regulated (not covered) garbage drums inside (fruit, food wraper, milk cartons were found inside). Found insect activity around wooden box.

Correct action required to avoid similar penalties: Receptacles used for the storage of garbage must be kept closed with a lid at all times. If cleaning of such receptacles is necessary, please carry out the cleaning after departure from the port and ensure that all receptacles are kept upright with lids closed after the cleaning.

3.4   Disposal Procedures

Although disposal is possible and consistent with revised Annex V, discharge of garbage to port reception facilities should be given the first priority. When disposing of garbage, the following points should be considered:

i) Disposal of un-compacted garbage is convenient but results in a maximum number of floating objects which may reach shore even when discharged beyond 25 miles from the nearest land. If necessary and possible, weights should be added to promote sinking. Compacted bales of garbage should be discharged in water depths of 50 meters or more to prevent breaking up from wave action and currents.

ii) Maintenance wastes contaminated with substances, such as oil or toxic chemicals, are in some cases controlled under other annexes or other pollution control laws. In such cases, the more stringent disposal requirements take precedence. Oily rags and contaminated rags must be kept on board and discharged to a port reception facility or incinerated.

iii) To ensure timely transfer of ship generated garbage to port reception facilities, ship agents are to be advised for guidance. Disposal needs should be identified particularly when arrangements are necessary for garbage requiring special handling.

iv) Cargo residue disposed at sea requires additional information to be recorded in the garbage record book (under Category G) such as discharge start and stop positions. Cargo residue should be collected in bags and disposed outside special areas and preferably more than 25 miles from shore or land at discharge ports in consultation with discharge port agents to avoid undue delay to vessel.

3.5   Summary of Disposal Requirements at Sea In and Outside Special Areas

According to the revised regulatory requirements discharges of all solid and hazardous waste into sea is prohibited except as given below.

For ALL types of vessels, outside Special Areas designated under Revised MARPOL Annex V:

- Comminuted or ground food wastes (capable of passing through a screen with openings no larger than 25 mm) may be discharged not less than 3 nautical miles from the nearest land
- Other food wastes may be discharged not less than 12 nautical miles from the nearest land
- Cargo residues classified as not harmful to the marine environment may be discharged not less than 12 nautical miles from the nearest land
- Cleaning agents or additives in cargo hold, deck and external surfaces washing water may be discharged only if they are not harmful to the marine environment.

With the exception of discharging cleaning agents in washing water, the ship must be en route and as far as practicable from the nearest land.

For ALL types of vessels, inside Special Areas designated under Revised MARPOL Annex V:

- More stringent discharge requirements apply for the discharges of food wastes and cargo residues, and
- As per shipboard waste management plan.

For fixed or floating platforms and All other ships when alongside or within 500 m of such platform

Exception: Comminuted or ground food wastes may be discharged from fixed or floating platforms located more than 12 miles from the nearest land and from all other ships when alongside or within 500 meters of such platforms.

Comminuted or ground food wastes must be capable of passing through a screen no larger than 25 millimetres.

Placard shown as Table 10 summarises the solid and hazardous waste disposal requirements, with colour code, laid down by the authorities for all ships and for the oil platforms. When garbage is mixed with or contaminated by other harmful and hazardous substances prohibited from discharges or having different discharge requirements, the more stringent requirements shall apply.

To note, special areas are the Mediterranean Sea, the Baltic Sea, the Black Sea, the Red Sea, The Gulf areas, North Sea, Antarctic area, and the wider Caribbean region including Gulf Of Mexico and the Caribbean Sea. Great Barrier Reef in Australia and Torres straights are also considered as special areas and no garbage is disposed in these special areas. Table 11 shows the adoption, entry into force and effective dates of these special areas for implementation in the revised Annex V.

3.5.1 Master's Overriding Authority

There are some exceptions for compliance of these regulations, when some unforeseen circumstances arise on board. In those special emergency conditions, Master has the authority to dispose of garbage outside the requirements of revised MARPOL Annex V. For instance,

- discharge of garbage from a ship is necessary for the purpose of securing the safety of a ship and those on board or saving life at sea, or
- accidental loss of garbage resulting from damage to a ship or its equipment, provided that all reasonable precautions have been taken before and after the occurrence of the damage, to prevent or minimize the accidental loss, or
- accidental loss of fishing gear from a ship provided that all reasonable precautions have been taken to prevent such loss, or

- discharge of fishing gear from a ship for the protection of the marine environment or for the safety of that ship or its crew.

In addition, an 'en route' exception included in Master' overriding authority, describes the 'en route' requirements, when discharging outside a special area and within a special area, does not apply to the discharge of food wastes, where it is clear that the retention of these food wastes on board a ship, presents an imminent health risk to the people living on board. If the Master disposes garbage in an exceptional circumstance, full records of ships position and nature and quantity of discharge shall be maintained in the garbage record book.

4.   Regulations for the Solid and Hazardous Waste Management System On Board

According to the revised MARPOL Annex V from 1st January 2013, all vessels should display placards, have a garbage management plan, and an approved garbage record book for proper record keeping of all types of discharges.

- Placards

a) Every ship of 12 metres or more in length overall and fixed or floating platforms shall display placards which notify the crew and passengers of the disposal requirements of regulations.

b) The placards shall be written in the official language of the State whose flag the ship is entitled to fly, and for ships engaged in voyages to ports or offshore terminals under the jurisdiction of other parties to the convention, in English or French.

- Garbage Management Plan

Every ship of 100 tons gross tonnage (GT) and above and every ship which is certified to carry 15 persons or more and fixed or floating platforms shall carry a garbage management plan which the crew shall follow. This plan provides written procedures for minimizing, collecting, storing, processing and disposing of garbage, including the use of the equipment on board. It shall also designate a person (s) in charge of carrying out the plan. Such a plan shall be in accordance with the guidelines developed by the organization and written in the working language of the crew and for all ships engaged in international voyages in English. A sample of solid and hazardous waste management plan of a car carrier ship is exhibited as Table 12 explaining the storage and disposal procedures for different categories of waste generated.

- Garbage Record Book

Every ship of 400 tons gross tonnage and above and every ship which is certified to carry 15 persons or more engaged in voyages to ports or offshore terminals under the jurisdiction of other parties to the convention and every fixed and floating platform shall be provided with a garbage record book. This book, whether as a part of the official logbook or otherwise, shall be in the form specified as shown in Appendix IV in this review article. In the discharge record book,

- Each discharge operation or to a reception facility, or completed incineration, shall be promptly recorded and signed for on the date of the incineration or discharge by the officer in charge. Each completed page shall be signed by the master of the ship. The entries in the book shall be at least in English, French or Spanish. The entries in an official national language of the flag the ship is entitled to fly shall prevail in case of a dispute or discrepancy,

- The entry for each incineration or discharge shall include date and time, position of the ship, category of the garbage and the estimated amount incinerated or discharged,

- The Garbage Record Book shall be kept on board the ship or the fixed or floating platform, and in such a place as to be available for inspection in a reasonable time. This document shall be preserved for a period of two years after the last entry is made,

- In the event of any discharge or accidental loss, referred to in regulation 6 of this Annex, an entry shall be made in the Garbage Record Book, or in the case of any ship of less than 400 gross tonnage, an entry shall be made in the ship's official logbook, of the location, circumstances of, and the reasons for the discharge or loss, details of the items discharged or lost, and the reasonable precautions taken to prevent or minimize such discharge or accidental loss.

- *The Administration may waive the requirements for Garbage Record Books for:*
  - any ship engaged on voyages of one hour or less in duration which is certified to carry 15 persons or more, or
  - fixed or floating platforms.

- *Inspection*

The competent authority of the Government of a Party to the Convention may inspect garbage record book or ship's official logbook on board any ship to which this regulation applies while the ship is in its port or offshore terminals and may make a copy of any entry in those books, and may require the master of the ship to certify that the copy is a true copy of such an entry. Any copy so made, which has been certified by the master of the ship as a true copy of an entry in the ship's Garbage Record Book or ship's official logbook, shall be admissible in any judicial proceedings as evidence of the facts stated in the entry. The inspection of the Garbage Record Book or ship's official logbook and the taking of a certified copy by the competent authority under this paragraph shall be performed as expeditiously as possible without causing the ship to be unduly delayed.

5.   Control of Solid and Hazardous Waste Management Plan

5.1   Designated Persons Responsible for Carrying Out Plan

The master has the overall responsibility for the implementation of the company's environmental protection policy. However, other designated personnel responsible for carrying out this plan on board is the chief officer assisted by chief engineer and other department heads.

The chief officer is the designated person who is responsible overall for carrying out the plan and record keeping. Being a technical head of engine department, the chief engineer is responsible for the overall management of all waste disposals devices and waste processing systems. The department heads are responsible for supervising the separation of garbage at source and removal of garbage to the designated area for processing, storage or disposal. For purpose of this plan the 2nd engineer and chief cook/cook are the in charge of their respective departments. The person in-charge may delegate the responsibility to others, who will work under his supervision and in compliance with the garbage management plan.

5.1.1   Duties of Team Members Who are Designated for Carrying Out Waste Management Plan

The Team members on board solid and hazardous waste management plan are chief officer, chief engineer, 2nd engineer and chief cook. Their duties are listed below:

Chief Officer

- Is overall responsible for carrying out the plan and record keeping
- If new deck crew joins the vessel, specific requirements relating to garbage management and its prohibition of disposal of plastics into the sea are explained
- Garbage management rules are followed on board as per the 'Waste Management Plan'
- Ensures all garbage bins are covered and retention area kept clean
- Ensures proper colour coding of garbage bins with stencils on them
- Makes non-regular staff and suppliers aware of environmental management systems (EMS) and consequences of departure from EMS procedures
- Ensures all packing materials from stores/spares supplied are landed back to supplier soon after storing.

Chief Engineer

- Is in charge of garbage management in engine room
- All garbage is segregated as per the 'Waste Management Plan'
- If any new engine crew join, then garbage management requirements are emphasised and made sure the crew understand the total prohibition of disposal of plastics into the sea

- No plastics are to be incinerated on board, all plastics to be collected and landed ashore
- No garbage is thrown overboard without checking with the Ch/Off.

2nd Engineer

- Assists Ch/Eng in garbage management plan in engine room
- Supervises the separation of garbage at source and removal of garbage to designated area for storage and disposal
- All garbage is to be segregated as per the 'Waste Management Plan'
- Ensures all garbage bins comply with colour code in engine room and stencilled
- Reports timings in consultation with Ch/Eng, when incinerator ash collected, time and quantity to be recorded in garbage log book by Ch/off
- Ensures all garbage bins are covered to stop spontaneous combustion in engine room
- No garbage to thrown overboard without checking with the Ch/Off.

Chief Cook

- In charge of garbage management in galley
- Supervise separation of garbage at source and removal of garbage to designated area for storage or disposal. All garbage to be segregated as per the 'Waste Management Plan'
- To liaise with Ch/Off and ensure all garbage management duties are complied with
- No plastics to be thrown overboard
- Ensure all garbage bins got cover in the pantry and galley
- No garbage is thrown overboard without checking with the Ch/Off.
- Before throwing food waste, permission must be obtained from the Ch/Officer.

FOR THE REFERENCE OF ALL CREW MEMBERS THE SPECIAL AREAS ARE

Mediterranean Sea Area, Baltic Sea Area, Black Sea Area, Red Sea Area, Gulfs Areas, North Sea Area, Antarctic Area and Wider Caribbean Region.

5.2.   Records

Appendix I and II present a typical garbage record book with various entries before and after revised MARPOL Annex V which came into force from 1st January 2013. Entries in the garbage record book are made on each of the following occasions:

When garbage is discharged to a reception facility ashore or to other ships:

- Date and time of discharge
- Port or facility, or name of ship
- Categories of garbage discharged
- Estimated amount discharged for each category in cubic metres

- Signature of officer in charge of the operation.

When garbage is incinerated:

- Date and time of start and stop of incineration
- Position of the ship (latitude and longitude) at the start and stop of incineration
- Categories of garbage incinerated
- Estimated amount incinerated in cubic metres
- Signature of the officer in charge of the operation.

When garbage is discharged into the sea in accordance with regulations 3, 4, 5 or 6 of Annex V of MARPOL[11]:

- Date and time of discharge
- Position of the ship (latitude and longitude). *Note*: for cargo residue discharges, include discharge start and stop positions.
- Category of garbage discharged
- Estimated amount discharged for each category in cubic metres
- Signature of the officer in charge of the operation.

Accidental or other exceptional discharges or loss of garbage into the sea, including in accordance with regulation 6 of Annex V of MARPOL:

- Date and time of occurrence
- Port or position of the ship at time of occurrence (latitude, longitude and water depth if known)
- Categories of garbage discharged or lost
- Estimated amount for each category in cubic metres
- The reason for the discharge or loss and general remarks.

5.2.1 Receipts / Certificates

The master obtains receipts / certificates from the receiving facility certifying the amount (estimated in m^3) and type (category) of garbage discharged ashore. Receipts or certificates are then kept together with the garbage record book. This book should be retained on board for two years after the date of last entry.

5.3 Crew Training and Education Program

When signing on for the first time on one of our vessels, all crew members receive an introduction to the company's environmental protection policy, the necessity to keep oceans free from all categories of solid and hazardous waste.

5.3.1 Facilitate Collection of Garbage

Collection, segregation and the treatment of garbage in keeping the oceans and surrounding waters clean is an integral part of all joining crew familiarisation programme. Commencing the voyage, ship's officers give a brief introduction to the voyage schedule at which all crew members as well as all other employees on board have to take part. All the crew members are also emphasised on the importance of this proper garbage treatment throughout the voyage. In addition, the garbage management plan is made available to everybody at any time for information.

5.3.2 Facilitate the Processing of Garbage

All persons who serve garbage processing devices must have sufficient knowledge of the respective plant. The designated person carries out trainings on board. The training programme contains particularly,

- the requirements of revised MARPOL Annex V
- the subdivision of garbage into categories
- the restrictions and ban of discharging garbage into the sea
- the instructions of this garbage management plan
- instructions or manuals for the processing equipment
- instructions for the inspection and maintenance of the appliances
- waste disposal requirements of the states and ports called.

5.3.3 Facilitate the Storage of Garbage

The designated person in charge of the plan carries out periodical training and education for facilitating the storing of garbage on board. All training programmes treat difficulties at the execution of this plan. The programme emphasises problems in implementing this plan and the program outcome is reported to the head office for reference to development of training and education programmes.

All crew, in particular members of the solid and hazardous waste management team should be familiar with the ship board solid and hazardous waste management plan and aware of their duties and responsibilities. Periodic training is arranged and all the crew members are familiarised with the process of solid and hazardous waste management with their respective roles. On board computer based training (CBT) is in place and courses like ISO 14001 Environmental Management, Introduction to MARPOL, Incinerator operations, National Pollutant Discharge Elimination system

(NPDES) designed by United States Environmental Protection Agency (EPA) are made compulsory for all the crew members to complete before they sign off from the ship. However, the success and effective implementation of waste management plan depends on three main factors, which is motivation from top level management (on board and ashore), motivation of the staff, and continuous training and involvement of the staff.

6   Solid and Hazardous Waste Management Flow Chart

Figure 3 illustrates the flow of solid and hazardous waste (garbage) from the point of generation, separation and disposal, whether it is for ocean disposal or for non-ocean disposal.

i.) Ocean disposable garbage - To be stored only for short term on board. It should be disposed at sea complying with the requirements as stated below or if shore reception facilities are available then it should be discharged ashore (especially if vessel is trading within the 12 nautical miles limit and en route).

ii.) Non-ocean disposable garbage - To be stored on board for long term if necessary and to be discharged to shore reception facilities at first opportunity.

iii.) When garbage is mixed with other harmful substances having different disposal or discharge requirements, the more stringent disposal requirements apply.

7   Discussions and Recommendations

Solid and hazardous waste, also referred as garbage, collection, storage, and processing by the use of equipment on board and disposal should be cost-effective and always provide environmentally sound results. However, on board the vessel, procedures in handling and storing garbage vary depending on factors such as the type and size of ship, the area of operation (eg. distance from land), ship's on board garbage processing equipment and storage space, crew size, duration of voyage, regulations and reception facilities at ports of call.

Therefore, for any solid and hazardous waste management plan to be effective, the design should be

- realistic, practical and easy to understand and use
- familiar to those with key functions on board the ship
- evaluated, reviewed and updated regularly, and
- tested regularly for its viability.

7.1   *Ship Board Management Philosophy Based on 'Upstream Solutions'*

All shipping companies should adopt a strategic philosophy which is based on 'upstream solutions' with respect to the environmental protection. This means that all people involved in operation of the

21

vessel should aim and work to prevent the 'creation of garbage generation' rather than 'controlling the generated garbage'.

It will also be economically advantageous to keep the garbage generation at a minimum level through a combination of complementary techniques and this are listed below.

i.) *Source Reduction:* The best means of reducing waste generated on board the vessel is to prevent waste generation from provisioning activities.

Following suggestions will reduce the volume of packing materials by suppliers.

- Immediately after receiving stores and provision on board cartons, box, pallets, or from any packing materials and reload to delivery track

- Prior delivery, the master can advise the chandler to minimize the utilization of packing materials

- This will be more effective if the purchasing department can co-ordinate to the chandelling company and the chandelling company to the supplier to minimize the utilization of packing materials

- However, it is very time consuming if unwrapping of stores and provision is to be carried out while receiving, so the best remedial action is to utilize majority of the crew, and

- The suppliers themselves must have their own environmental programs and procedures as to how to handle the stores, provisions, and any product with less packing materials but retain its good quality and condition.

In line with the strategy of 'prevent creation of waste rather than controlling later,' there has been a complete ban of using 'asbestos' a hazardous and harmful substance, in new installations on both new and existing convention ships, since 1st January 2011[12].

ii.) *Recycling or reusing:* This is one of the options in minimizing the waste generated. How ever, if this option is attempted, due consideration must be given to the limited storage space available at the designated storage stations, shipping companies mandatory requirements for good house keeping with regards to hygiene, safety. In other words, this option, never, should lead to fire risk and creating any sort of pollution threat. As per the amended regulations, recyclable waste has be identified and stored in green receptacles and then discharged to shore reception at first available opportunity.

iii.) *Disposal:* Disposal to shore port reception facilities thus preventing the pollution of sea by garbage generated.

In view of the costs involved with the different ultimate disposal techniques, it may also be economical and advantageous to keep garbage requiring special handling (eg. hazardous wastes) separate from other garbage collection.

More importantly, priority should always be given for 'disposal ashore to the port reception facilities'. However, some notification procedures should be laid down and in place while disposing the waste generated on board the ship.

7.2 *Disposal to Port Reception Facilities and Disposal Notifications of Solid and Hazardous Waste*

IMO has recognized that provision of adequate reception facilities is crucial for effective MARPOL implementation. The marine environment protection committee (MEPC), is also strongly encouraging all member states, particularly those parties to the MARPOL Convention as Port states, to fulfil their treaty obligation on providing adequate port reception facilities. Further, MEPC emphasized the importance of port reception facilities in the chain of implementation of the MARPOL convention and stated that the policy of 'zero tolerance of illegal discharges from ships',[13] whether it is oil or solid or hazardous waste, can be effectively enforced when there are adequate reception facilities in all ports.

For the ship's Master and crew, two disposal notification forms are available as per the revised Annex V. Form 1 incorporates the advance notification for solid and hazardous waste disposal to shore reception facilities and is in a standard format, whilst form 2 highlights reporting inadequacies of the port reception facilities.

According to the form 1, the Master shall provide advance notification of the type and quantity of garbage on board to be disposed to the designated authority at least 24 hours in advance of arrival or upon departure of the previous port if the voyage is less than 24 hours. A typical format of this reporting form is exhibited as Appendix III. This form shall be retained on board along with the garbage record book.

The Master is also obliged under MARPOL Annex V requirements, to report any non-availability of sufficient reception facilities at any port using the IMO standard format, and form 2 which will be readily available in the official garbage record book of the company, as shown in Appendix IV.

With regard to the need and requirement of reporting inadequate port reception facilities, there is already an action plan in place,[14] by the MEPC in tackling the alleged inadequate port reception facilities which is seen as major hurdle to overcome in achieving full compliance with MARPOL. This plan includes,

- work items related to reporting requirements
- provision of information on port reception facilities
- identification of any technical problems encountered during the transfer of waste from ship to shore
- the standardization of garbage segregation requirements and containment identification
- review of the type and amount of waste generated on board and the type and capacity of port reception facilities
- revision of the IMO Comprehensive Manual on Port Reception facilities, and
- development of guide to Good Practises on Port Reception Facilities.

To summarize, MARPOL imposes numerous operational and technical requirements on ships and at the same time, MARPOL also enforces one important obligation to the Government of each Party to the Convention, which is to provide facilities for the reception of ship generated residues and garbage (solid and hazardous waste) that cannot be discharged into the sea. The reception facilities must be adequate to meet the needs of ships using the port, without causing undue delay to ships operation. Thus, the requirements for port reception facilities create an incentive for ships to comply with MAPOL and minimize the discharges to sea.

7.3   *Solid and Hazardous Waste Management Audits*
One way of testing the success or efficiency of this solid and hazardous waste management plan implementation is by setting up audits. Regular auditing and quantification of wastes generated are therefore, required on board. For instance, a regular assessment of at least once a year of the different categories of waste generated, will provide an opportunity to analyse waste-related issues. Identification of significant quantities/amounts of wastes produced, could also be assessed to reduce the amount and costs of garbage disposal. A well prepared solid and hazardous waste management plan will thus serve as a base for assessment. For a better understanding of this concept, a garbage data sheet of a car carrier ship for the year ending 2012, used in waste management auditing is elucidated in Table 13.

24

7.3.1 Solid and Hazardous Waste Management is an Integral Part of Shipping Company's Environmental Management System

Solid and hazardous waste management plan is an integral part of the company's environmental protection policy and should be implemented in line with the ship's specific environmental management program. Figure 4 illustrates the flow of information from management office to the ships and vice versa, in implementing the shipping company's environmental protection policy. This process includes identification of various environmental aspects, setting their objectives and targets, suggested measures to take for achieving these targets, periodical review of their status and naming the person responsible, on board/ashore, to achieve acceptable progress to reduce their impacts on the environment. Table 14 shows an example of an ocean going car carrier vessel's 'ship specific environmental program with targets and status' identifying different environmental aspects which can possibly cause environmental impact to the environment such as to the air, or to the sea or other impacts.

One of the identified environmental aspects is solid/dry and hazardous waste handling, management and disposal. For example, for plastics which are considered as a hazardous waste, the objective is to minimize or avoid using plastic and plastic related material with a set target of reduction consumption level by 5% based on the accumulated record for the past 12 months. In addition, to lower/maintain this reduction, the ship is also following the practise of returning plastic packing to the supplier / ship chandler while receiving stores / spares.

Table 15 provides the objectives with a target of reduction of the volume of the garbage generated by 5% or more compared to the previous year 2012, derived from the ship specific environmental program for that particular year. In this table, as against plastic based on the previous year quantity of plastic disposed ashore, the target set was 35 m3 for the year 2012. It is quite clear that the targets were met as the quantity disposed to shore was only 32.5 m3 in year 2012. The target set for the current year 2013 is 30 m3, which is again about 5% less consumption than the previous year. Further, there is a also a 'planned target' of 5% reduction of the waste disposal to port reception facilities for year 2013 as described in Table 16.

Similar targets have been set for other environmental aspects in the ship specific environmental program as well, and every crew member should be briefed about these targets in the environmental committee meetings. All people involved in the garbage management plan should try to achieve these targets. The best measures in achieving these targets are:

- selection of suppliers whose products minimise waste

- selection of suppliers who provide a refill or replacement service for their products
- purchase of products in larger packing units
- removal of excess packaging before loading
- returning old and out of date unused equipment (eg. time expired pyrotechnics and pharmaceuticals) to the manufacturers or sellers for reuse, recycling, refilling, or disposal as appropriate
- complete emptying of containers and full use of the contents
- optimisation of waste producing processes so as to reduce waste
- evaluation of product-specific waste quantities to identify means to reduce waste
- use of products with a long lifespan and / or shelf life
- use of equipment that can be repaired, and
- repair rather than replace equipment

To optimise this solid and hazardous waste management practises on board, all crew members should be actively engaged in the implementation process, since the success of any measure put in place depends to a larger extent on crew's acceptance and involvement.

7.4 *Classification Societies Initiative Regarding Revised Annex V*

In light of these new requirements coming into force, the ship classification societies, like DNV, are introducing a new 'Statement of Compliance' with respect to Annex V (garbage pollution prevention). This statement of compliance confirms that the equipment on board the ship is in accordance with the new revised Annex V with reference to garbage management plan, placards, garbage record book, containers, processing equipment etc.). The statement of compliance has to be issued after a successful initial survey and should be renewed on every renewal survey (subject to five-yearly survey scheme). Since Annex V does not prescribe any sort of certification, the above initiative of Statement of Compliance by the classification societies, will help the owners to demonstrate compliance to any interested parties, such as port state control or labour authorities etc.

8.   Final Word

To obtain the most efficient way of managing solid and hazardous waste, and to reduce burden in segregating and handling this on board the ship, as well as on shore in the ports, the concept of waste minimization has been introduced. This should be given highest priority by following the basic principle of 'avoidance before reduction before recycling before disposal'.[15]

The responsibility for avoiding discharges like, solid and hazardous waste, for that matter any harmful substances, rests not only with the master and his crew but also with all other stake holders in the trade such as with the chatterer, the ship owner and the port authorities.

The Master and the crew on board a vessel should be fully proficient in carrying out the correct procedures and should apply them carefully and consciously. The chaterer should also include in the Charter Party a clause stating his policy on pollution prevention compliance. The ship owner on the other hand, should ensure sound management practises in safety and pollution prevention, as required by the International Safety Management Code on board his ship which is, in line with the company's environmental prevention policy. The ports and port authorities must be prepared to accept and provide adequate reception facilities to all categories of solid, hazardous and harmful waste generated by the ships in operation, without causing undue delay to ships.

These four main stake holders are the key persons who can actually make the vision of 'zero discharges' to the sea a reality, and in keeping the world oceans and beaches clean.

9.   References

1. International Chamber of Shipping and International Shipping Federation. Newsletter, May 2008

2. European Maritime Safety Agency – Equasis Statistics: The World Merchant fleet in 2011.

3. UNCTAD (United Nations Conference on Trade and Development) Review of maritime transport 2007, Report by the UNCTAD secretariat, United Nations, New York and Geneva 2007, page 167.

4. Hellenic Marine Protection Association (HELMEPA).

5. International Maritime Organization (ISM) Code - International safety management code which is mandatory for all ships and shipping companies, is for protecting people, property and environment by reducing the risk of operation and plus ensuring continuous improvement of operation through the dynamic elements of the code.

6. International Maritime Organization, Revised MARPOL Annex V (including amendments) Regulations for the Prevention of Pollution by Garbage from Ships, MARPOL   Consolidated edition 2006 Articles, Protocols, `, Unified Interpretations of the International Convention for the Prevention of Pollution from Ships, 1973, as modified by the Protocol of 1978 relating there to, MARPOL Published by IMO London 2006, Page 322.

7. International convention for the Prevention of Pollution from ships, 1973/78 MARPOL, is the regulations covering the various sources of ship generated pollutions are contained in six annexes of the convention. They are Annex I Regulations for the prevention of pollution by oil, Annex II

Regulations for the control of pollution by noxious liquid substances in bulk. Annex III Regulations for the prevention of pollution by Harmful substances, Annex IV Regulations for the prevention of Pollution by sewage from ships, and MARPOL Annex VI Regulations for Prevention of Air pollution from ships.

8. MARPOL 73/78 Annex VI Regulation for Prevention of Pollution from ships and Appendix IV Type Approval and Operating Limits for Ship Board Incinerators (regulation 16)

9. State of Flag: A ships' flag state exercises regulatory control over the vessel and is required to inspect it regularly certify the ship's equipment and crew and issue safety and pollution prevention certificates / documentation periodically.

10. DNV Technical e-News letter, Recording of Incinerator Capacity on the Supplement to the IOPP certificate, Page 1, dated 17 January 2012.

11. Revised MARPOL Annex V, Regulations 3: Disposal of garbage outside special areas, Regulation 4: Special requirements for disposal garbage, Regulation 5: Disposal of garbage within special areas, Regulation 6: Exceptions to regulations 3, 4 and 5 and Regulation 7: Port reception facilities. Further Regulation 8: Port state Control on operational requirement, and Regulation 9: Placards, garbage management plans and garbage record-keeping

12. DNV Technical e Newsletter dated 29th March 2012, DNV follow-up of the asbestos ban.

13. 54th Session of MEPC March 2006.

14. Nikos Mikelis IMO's presentation paper on 'Ships' Waste: Time for Action' in IMO's action plan on Tackling the inadequacy of port reception facilities, Organized by EUROSHORE and FEBEM-FEGE Supported by OVAM Brussels, October14, 2010.

15. ISO 21070 Ships and marine technology-marine environment protection-Management and handling of ship board garbage, Introduction page vi, ISO 2006.

Further reading regulatory requirements are:

- Revised MARPOL Annex V MEPC.201 (62) adopted 15th July 2011
- IMO guidelines on the implementation of Annex V of MAPOL 73/78 MEPC.219 (63) adopted 2nd March 2012
- IMO Guidelines for the development of Garbage Management Plans MEPC.220 (63) adopted 2nd March 2012
- ISO 21070 'Standard for the management and handling of ship board garbage'.

Table 1. World Fleet: Total Number of Ships, by Type and Size Year Ending 2011.

Ship Type	Small GT <500	Medium 500 GT < 25000	Large 25000 GT < 60000	Very Large > GT 60000	Total
General cargo ships	4,627	12,210	197		17,034
Specialized cargo ships	14	188	48		250
Container ships	16	2,411	1,679	868	4,974
Ro-Ro cargo ships	32	774	587	144	1,537
Bulk carriers	362	3,647	4,215	1,373	9,597
Oil and Chemical tankers	1,852	6,373	2,255	1,348	11,828
Gas tankers	44	1,014	187	329	1,574
Other tankers	259	402	5		666
Passenger ships	3,461	2,505	269	135	6,370
Offshore vessels	2,185	4,312	75	120	6,692
Service ships	2,196	2,219	23	4	4,442
Tugs	13,238	872			14,110
GRAND TOTAL	28,286	36,927	9,540	4,321	79,074

Source: European Maritime Safety Agency-Equasis Statistics of the World Merchant fleet in 2011.

GT: Gross tonnage – refers to the volume of all ship's enclosed spaces, from keel to funnel, measured to the outside of the hull framing

Table 2. Volume of the Solid and Hazardous Waste (also Referred as Garbage) in M^3 Generated Per Year in a Typical Ocean Going Car Carrier Ship.

SNo	Garbage Categories	Disposed to Shore Reception Facilities (m^3)	Incinerated On Board (m^3)	Disposed at Sea (m3)	Total Volume Produced in a Year (January 2012 to December 2012) (m^3)
1	Plastic (hazardous waste)	32.5	none	none	32.5
2	Floating dunnage, lining, or packing material	0.0	none	0.0	0.0
3	Paper products, rags, glass, metal, bottle and crockery	none	none	1.96	1.96
4	Food waste	none	none	3.33	3.33
5	Incinerated ash	0.05	none	none	0.05
6	Used batteries (hazardous waste)	222 pcs (0.122)	none	none	0.122
7	Aerosol and similar sprays	0.04	none	none	0.04
8	Used florescent tubes and bulbs (hazardous waste)	725 pcs (1.33)	none	none	725 (1.33)
9	Others like oily waste and oily rags etc. (hazardous waste)	41.05	5.42	none	42.6
	GRAND TOTAL in m3	75.1	5.42	5.29	85.81

Note: Measures taken to minimize the garbage quantity at source by returning lining and packing materials to the supplier.

Table 3. Time Taken for Different Objects to Dissolve at Sea.

General Object	Time Taken to Dissolve at Sea
Paper bus ticket	2-4 weeks
Cotton cloth	1-5 months
Rope	3-14 months
Woolen cloth	1 year
Painted wood	13 years
Tin can	100years
Aluminum can	200-500 years
Plastic bottle	450 years

Source: Hellenic Maritime Environmental Protection Association (HELMEPA).

Table 4. Areas of Responsibility for Collection and Separation of Solid and Hazardous Waste On Board Ocean Going Vessels.

Area	Head of Department Responsible	Person Responsible for Management of Collection and Separation
Cargo holds / Deck dept. (including working spaces and store rooms in accommodation)	Chief officer	Bosun / AB's / Ordinary seaman
Engine dept. (including working space and store rooms in accommodation)	Chief engineer	2nd Engineer / Oilers / Motorman / Wipers
Medical locker / hospital	Chief officer	2nd Officer / Deck cadet
Galleys / pantries / officer & crew dining / officers Saloon and mess rooms / library / conference room	Head of catering department	Chief Cook / Cook / Mess man
Public wash rooms	Head of catering department	Cook / Mess man
Spa / Gym	Head of catering department	Cook / Mess man
Laundry room	Head of catering department	Cook / Mess man
Officer / Crew alleyways	Head of the department	Shared between deck, engine and catering dept. crew

Table 5. Various Equipment and Processing of Solid and Hazardous Waste On board.

Type of Equipment	Location	Wastes Processed	Processed Waste
Galley Comminuter or Grinder	Galleys	Food waste	Into sea and / or dried and incinerated
Shredder Machine	Engine room	Bottles, glass, crockery	Bagged and stored until landed ashore
Bale Compactor	Engine room	Cardboard and paper	Bundled and stored until landed ashore or incinerated
Densifier	Engine room	Tins, cans, metal	Bundled and stored until landed ashore
Garbage Shredders	Engine room	Burnable waste	Incinerated
Incinerators	Engine room	Burnable waste	Ash bagged and stored until landed ashore

Table 6. Importance of Compactor in Solid and Hazardous Waste Management Plan.

Examples of Garbage	Special Handling by Vessel Personnel before Compaction	Rate of Alteration	Retainment of Compacted Form	Density of Compacted Form	Onboard Storage Space
Metal, food and beverage containers, glass, small wood pieces	None	Very rapid	Almost 100%	High	Minimum
Comminuted plastics, fibre and paper board	Minor – reduce material to size for feed, minimal manual labour	Rapid	Approximately 80%	Medium	Minimum
Small metal drums, uncomminuted cargo packing, large pieces of wood	Moderate – longer manual labour time required to size material for feed	Slow	Approximately 50%	Relatively low	Moderate
Uncomminuted plastics	Major – very long manual labour time to size material for feed; usually impractical	Very slow	Less than 10%	Very low	Maximum
Bulky metal cargo containers, thick metal items	Impractical for shipboard compaction; not feasible	Not applicable	Not applicable	Not applicable	Maximum

Compactors make generated waste easier to store and easier to transport to shore reception facilities.

Table 7. Incinerator Characteristics and Procedures for Shipboard Generated Waste.

Typical examples	Handling by Vessel Personnel	Combus-tibility	Reduction of Volume	Residual	Exhaust	On Board Storage Space
Paper packing food and beverage containers	Minor-easy to feed into	High	Over 95%	Powder ash	Possibly smoky and not hazardous	Minimum
Fibre and paper board	Minor-reduce material to size for feed, minimal manual labour	High	Over 95%	Powder ash	Possibly smoky and not hazardous	Minimum
Plastic packaging, food and beverage containers	Minor-easy to feed into hopper	High	Over 95%	Powder ash	Possibly smoky and hazardous based on incinerator design	Minimum
Plastic sheeting, netting, rope and bulk material	Moderate manual labour time for size reduction	High	Over 95%	Powder ash	Possibly smoky and hazardous based on incinerator design	Minimum
Rubber hoses and bulk pieces	Major manual labour time for size reduction	High	Over 95%	Powder ash	Possibly smoky and not hazardous based on incinerator design	Minimum
Metal food and beverage containers	Minor-easy to feed into hopper	Low	Less than 10%	Slag	Possibly smoky and not hazardous	Moderate
Metal cargo, bulky containers, thick metal items	Major manual labour time for size reduction (not easily incinerated)	Very Low	Less than 5%	Large metal fragments and slag	Possibly smoky and not hazardous	Maximum
Glass food and Beverage containers	Minor-easy to feed into hopper	Low	Less than 10%	Slag	Possibly smoke and not hazardous	Moderate
Wood, cargo containers and cargo wood scrapes	Moderate manual High labour time for size reduction	High	Over 95%	Powder ash	Possibly smoky and not hazardous	Minimum

Residue from plastics like hazardous material incineration is still considered as hazardous and thus cannot be discharged into the sea.

Table 8. Overview of Management of all Types of Solid and Hazardous Water Generated On Board an Ocean Going Vessel.

Garbage Type	Treatment	Documentation	Disposal/Storage
Plastics	Incinerator	Garbage record Book/Receipts	Storage till disposal ashore
Oily rags and waste	Incinerator	Garbage record Book/Receipts	Storage till disposal ashore
Floating dunnage, lining or packing material	Bale Compactor Incinerator	Garbage record Book/Receipts	Storage till disposal ashore
Paper products, cardboard, non-oily rags	Bale Compactor Incinerator	Garbage record Book	Storage till disposal ashore
Food waste	Ground / comminuted	Garbage record Book	Storage till permitted disposal at sea
Glass, crockery	Shredder	Garbage record Book/Receipts	Storage till disposal ashore
Aluminium and metal cans, metal	Densifier	Garbage record Book/Receipts	Storage till disposal ashore
Galley grease and used cooking oil	Incinerator	Garbage record Book	Storage till disposal ashore
Dry Cleaning waste fluids	Hazardous waste - Storage	Garbage record Book/Receipts	Storage till disposal ashore
Paint and related Wastes	Hazardous waste - Storage	Garbage record Book/Receipts	Storage till disposal ashore
Other Chemical waste	Hazardous waste - Storage	Garbage record Book/Receipts	Storage till disposal ashore
Photocopier and Laser printer cartridges	Hazardous waste - Storage	Garbage record Book/Receipts	Storage till disposal ashore
Fluorescent, mercury vapour light bulbs & thermometers	Hazardous waste - Storage	Garbage record Book/Receipts	Storage till disposal ashore
Lead acid batteries (LAB)	Hazardous waste - Storage	Garbage record Book/Receipts	Storage till disposal ashore
Nickel-cadmium, Lithium, alkaline batteries.	Hazardous waste - Storage	Garbage record Book/Receipts	Storage till disposal ashore
Pesticides	Hazardous waste - Storage	Garbage record Book/Receipts	Storage till disposal ashore
Smoke Detectors	Hazardous waste - Storage	Garbage record Book/Receipts	Storage till disposal ashore
Incinerator ash	Land ashore	Garbage record Book/Receipts	Storage till disposal ashore.
Bio-hazardous	Directly to	Garbage record	-

	Incinerator	Book	
Pyrotechnics	Used or soaked in water require only normal disposal. If expired, return to supplier	Garbage record Book/Receipts	Storage till disposal ashore
Hypodermic needles.	Return to supplier	Garbage record Book/Receipts	Storage in medical centre till disposal to approved company ashore
Pharmaceuticals (PHA)	Return to supplier or as per flag state requirements	Garbage record Book/Receipts	Storage in medical centre till disposal ashore
Cargo residues (not harmful to marine environment)	If possible land ashore at discharge port or dispose at sea	Garbage record Book/ Receipts	Storage till disposal at sea / ashore
Cargo residues (harmful to marine environment)	Hazardous waste - Storage	Garbage record Book/Receipts	Dispose ashore

Table 9. Segregation and Storing of Solid and Hazardous Waste in Marked / Coloured Bins For Disposal at Sea or Discharge to Shore Port Reception Facilities.

Disposal at Sea	Colour Code	Material
NO	RED	Non-recyclable plastics and plastics mixed with non-plastic garbage.
NO	BLACK	Rags
NO	GREEN	Recyclable material
YES	BLUE	Food waste
NO	YELLOW	Floating dunnage, lining etc.
NO	GREY	Oily rags, light bulbs, acids, chemical, batteries, etc.

Note: Grey bin: Items, which are hazardous to the environment and personnel, should not be mixed, rather individual waste streams are kept separate.

Table 10. Placard of Summary of Solid and Hazardous Waste Disposal Requirements at Sea and Outside Special Areas as per the revised MARPOL Annex V Resolution MEPC 201 (62).

Type of garbage[1]	Ships outside special areas	Ships within special areas	Offshore platforms (more than 12 nm from land) and all ships within 500 m of such platforms[4]
Food waste comminuted[2] or ground	Discharge permitted >3 nm from the nearest land as far as practicable and en route	Discharge permitted >12 nm from the nearest land as far as practicable and en route[3]	Discharge permitted
Food waste not comminuted[2] or ground	Discharge permitted >12 nm from the nearest land as far as practicable and en route	Discharge prohibited	Discharge prohibited
Cargo residues[5,6] not contained in wash water	Discharge permitted >12 nm from the nearest land as far as practicable and en route	Discharge prohibited	Discharge prohibited
Cargo residues[5,6] contained in wash water		Discharge permitted >12 nm from the nearest land as far as practicable, en route and subject to two additional conditions[7]	Discharge prohibited
Cleaning agents and additives[6] contained in cargo hold wash water	Discharge permitted	Discharge permitted >12 nm from the nearest land as far as practicable, en route and subject to two additional conditions[7]	Discharge prohibited
Cleaning agents and additives[6] in deck and external surfaces wash water		Discharge permitted	Discharge prohibited
Carcasses of animals carried on board as cargo and which died during the voyage	Discharge permitted as far from the nearest land as possible and en route. Should be >100nm and max water depth	Discharge prohibited	Discharge prohibited
All other garbage including plastics, synthetic ropes, fishing gear, plastic garbage bags, incinerator ashes, clinkers, cooking oil, floating dunnage, lining and packing materials, paper, rags, glass, metal, bottles, crockery and similar refuse	Discharge prohibited	Discharge prohibited	Discharge prohib

1 When garbage is mixed with or contaminated by other harmful substances prohibited from discharge or having different discharge requirements, the more stringent requirements shall apply

2 Comminuted or ground food wastes must be able to pass through a screen with mesh no larger than 25 mm.

3 The discharge of introduced avian products in the Antarctic area is not permitted unless incinerated, autoclaved or otherwise treated to be made sterile.

4 Offshore platforms located 12 nm from nearest land and associated ships include all fixed or floating platforms engaged in exploration or exploitation or associated processing of seabed mineral resources, and all ships alongside or within 500 m of such platforms.
5 Cargo residues means only those cargo residues that cannot be recovered using commonly available methods for unloading.
6 These substances must not be harmful to the marine environment.
7 According to regulation 6.1.2 of MARPOL Annex V the discharge shall only be allowed if: (a) both the port of departure and the next port of destination are within the special area and the ship will not transit outside the special area between these ports (regulation 6.1.2.2); and (b) if no adequate reception facilities are available at those ports (regulation 6.1.2.3).

Table 11. Summary Adoption, Entry into Force and Effective Dates of Special Areas.

Special Areas	Adopted	Date of Entry into Force	In Effect from
Mediterranean sea	2nd Nov 1973	31st Dec 1988	01st May 2009
Baltic sea	2nd Nov 1973	31st Dec 1988	1st Oct 1989
Black sea	2nd Nov 1973	31st Dec 1988	**
Red sea	2nd Nov 1973	31st Dec 1988	**
Gulf Sea	2nd Nov 1973	31st Dec 1988	01st Aug 2008
North sea	17th Oct 1989	18th Feb 1991	18th Feb 1991
Atlantic area (south of latitude 60 degrees south)	16th Nov 1990	17th Mar 1992	17th Mar 1992
Wider Caribbean region including the Gulf of Mexico and the Caribbean sea	4th July 1991	4th April 1993	01st May 2011

Table 12. Solid and Hazardous Waste Management Plan.

	VESSEL:	

STORAGE	DISPOSAL	

STORE IN RED RECEPTACLES
NON-OCEAN DISPOSABLES

Incinerable Garbage: Plastics, styrofoam, oily sludge, oily waste, oily rags, used cooking oil.
Non-Incinerable Garbage: Chemicals, paint drums, paint scrapings, pyrotechnics, medicines, aerosol cans, fluorescent tubes, batteries, PVC plastics, impregnated woods.

Waste that can be incinerated should be burnt at first opportunity, and the ashes kept for disposal ashore. If shore disposal is impossible, and the ship is not in a **Special Area**, ashes from wastes other than plastic may be dumped at sea outside the 12 n.m. limit and en route

IF IN DOUBT OF PLASTIC, DO NOT DUMP

Waste not recommended for incineration should be stowed for shore disposal.
Pyrotechnics require special disposal ashore.

GARBAGE MANAGEMENT TEAM

Job Description for the Team:
- Collection of garbage
- Sorting of garbage
- Storage of garbage
- Operating comminutor or grinder
- Operating incinerator
- Disposal of garbage and ashes
- Record keeping

Members of the Team:

Ch.Officer _____

Ch.Engineer _____

Ch.Cook _____

2ⁿᵈ Engineer _____

STORE IN GREY RECEPTACLES
NON-OCEAN DISPOSABLES

Oily rags.
Company require shore disposal

⇨

STORE IN GREY RECEPTACLES
NON-OCEAN DISPOSABLES

Oily rags.
Company require shore disposal

Notes:
When sorting waste, if plastic waste is mixed with other waste, then whole should be treated as plastic. Operators of comminutor or incinerator must have proper training in the operation of such machinery

---

STORE IN BLUE RECEPTACLES

Food waste
Company recommend shore disposal

⇨

Disposal at sea permitted for comminuted food waste (diam. <25mm) outside 3 n.m. limit; non-comminuted food waste outside 12 n.m. Limit and en route.
In **Special Areas** food waste may be disposed at sea outside 12 n.m. limit only and en route

DESIGNATED STORAGE AREA

DK, 7, Incinerator Room. Upper Deck Port Side
Note: All garbage shall be stowed in sound, securely covered containers.

STORE IN BLACK RECEPTACLES

Paper, rags, metal, glass, crockery, soot.
Company require shore disposal if the vessel's trade permits

⇨

Disposal at sea permitted for comminuted or ground waste outside 3 n.m. limit, non-comminuted waste outside 12 n.m. limit only and en route.
In **Special Areas** no disposal at sea allowed.

STORE IN YELLOW RECEPTACLES

Floating dunnage, lining, packing material.
Company require shore disposal if the vessel's trade permits

⇨

Disposal at sea permitted, but only outside the 25 n.m. limit and en route.
In **Special Areas** no disposal at sea allowed.

Authorisation:
Signature _____
Date _____

STORE IN GREEN RECEPTACLES

Recyclable materials.

⇨

STORE IN GREEN RECEPTACLES

Disposal at sea not permitted.

Table 13. An Example of Garbage Data Sheet for Use in Waste Management Auditing.

MONTH	DISCHARGED INTO THE SEA		DISCHARGED TO RECEPTION FACILITIES (m³)			INCINERATED (m³)	TOTAL IN MONTH
	CAT-4/3 (m³)	CAT-5 (m³)	CAT-1 (m³)	OTHERS (m³)	PLACE & DATE	CAT-4 (m³)	
JAN.	0.06	0.3				0.4	0.76
FEB.		0.13	4	6	Ulsan 16-Feb-2012 Xingang 23-Feb-2012	0.1	10.23
MAR.		0.24	2.00	4.00	Aqaba 18-Feb-2012	0.10	6.34
APR.		0.27	3	5.0	Ulsan 08-Apr-2012	0.60	8.87
MAY		0.25	4.00	6	Aqaba 06-May-2012 Shangahai 27-May-2012	0.30	10.55
JUN.	0.5	0.34				1.50	2.34
JUL.	0.15	0.30	4.00	6.2	Xingang 17-Jul-2012	1.02	11.67
AUG	0.45	0.32	1.50	2.0	Durban 13-Aug-2012	0.8	5.07
SEP		0.3	4.0	4.0	Ulsan 23-Sept-2012	0.3	8.6
OCT		0.31	3.0	3.0	Aqaba 16 Oct-2012		6.31
NOV	0.8	0.3	2.00	3	Ulsan 14th Nov-2012	0.10	6.2
DEC		0.27	5.00	3.40	Ulsan 26th Dec-2012	0.20	8.87
TOTAL	1.96	3.33	32.5	42.6		5.42	85.81

42

Table 14. A Typical Ship Specific Environmental Program With Targets and Status for Year 2013.

SHIP'S SPECIFIC ENVIRONMENTAL PROGRAM					Valid From: Jan 2013 - Dec 2013	Revision: 01	Rev. Date: 01 Jan'13 for end 2012	Page : 01 of 01
						Prepared By: Chief Officer	Approved By: Master-	
SNo	Aspects	EMISSION IMPACTS			Objective	Target	Measures	Accptable/ Non-Acceptable progress	Status	Responsible Officer / Target Date
		Air	Sea	Other						
1	Chemicals & Detergents		X		Minimize Consumption & on going studies.	To reduce consumption by 5% based on the accumulated record of the past 12 months.	1.Close monitoring of consumption. 2. Proper amount of detergents & chemicals to be used. 3. Boiler and cooling water analysis to be kept within limits but at a lower level	Acceptable	1. Monthly inventory of aspects. 2. Diluting chemicals with water & right ratio of detergent. 3. To inquire/updates with Supplier.	1.Cheng/Choff/ Chief Cook 2. Bosun, Messman 3. Office/Cheng / Choff / Chief Cook (Dec. 30, 2013)
2	Batteries/ Flourescent tubes		X		Minimize consumption of disposable batteries and to extend life expectancy of flourescent tubes.	To reduce consumption by 5 % based on accumulated record of the past 12 months	Retained/segregated for shore disposal. Rechargable batteries to be used.	Acceptable	1. Batteries consumed - 222 pcs which is 18.5pc/month. Target has been achieved for batteries. Set target for 2013 - 19pcs/month 2. A total of 725 pcs of f-tubes were disposed on shore facilities as per ship's record. An Average monthly consumption of 60.4pcs. were used/consumed for the year 2012. Target for 2012 set at 65 pcs, so vessel is well within target. Target for Year 2013 set at 58pcs/month	Electrician (Dec. 30, 2013)
3	Plastics	X	X	X	Minimize or avoid using plastic and plastic related materials. To reduce plastic consumption by 5 %	1.To reduce consumption of plastics by 5 % based on the accumulated record for the past 12 months. 2. To return plastic packings to supplier when receiving stores. 3. No disposal of plastic.	1. Minimize consumption of plastics. 2. Sent back to supplier 3. Re use chemical plastic container for Sludge and Cooking oil collection.	Acceptable	1 In progress, to monitor previous consumption and ensure target for the next year is meet. 2. Target for 2013 is 30 m³ - Disposed in 2012 - 32.5 cbm. Target for 2012 was 35m3 (Target Met)	Chief Officer / Fourth Engineer (Dec. 30, 2013)

.....contd

43

#	Item				Objective	Description / Action	Status	Remarks	Responsible (Target Date)
4	Papers	X	X	X	Minimize Consumption	To reduce consumption by 5 % based on the accumulated record for the past 12 months (2012). Paper (stationery) present year target set at 1000 pcs. of paper (stationery) per monthly consumption.	1.To use reverse side of used papers for printing internal, non-official documents. 2.Control unnecessary copying. 3.Monthly updating of paper supplies.	Acceptable	1. Practiced onboard 2. In progress 3. Monthly inventory carried out	1. All 2. All 3.Radio Officer & All Department Heads (Dec. 30, 2013)
5	Antifouling Paint		X		Undergoing studies	Undergoing studies	Suggest to order environmentally friendly products (e.g. tin-free paints)	Acceptable	5 yearly interval of dry docking was set by company.	Vessel Manager (Through Master)
6	Paint		X		Maintain Consumption	1. To maintain present consumption based on the accumulated record for the past 12 months. 2. Present year target set at 300L of paint consumption per month.	1.Paint cans manually compacted onboard/retained for shore disposal. 2. Proper planning of purchase (Good for 3 months interval)	Acceptable	1.Proper planning of PMS were carried out on board for efficient application of paint. 2.A total of 262L (average monthly consumption)was used in 2012, compared to 275Ls for 2011. Target achieved	Ch Off / Bosun (Dec. 30, 2013)
7	Sludge	X	X		Reduced production of sludge	Average production of sludge 16m3 per month	1. Run at economical Speed. 2. Incinerate 3. Reception facility ashore.	Acceptable	Incinerated (Jan - Dec 12): 55 m3	Chief Engineer (Dec. 30, 2013)
8	Ballast		X		Eliminate micro organisms in ballast water.	Clean Ballast onboard.	1. Plan ballast for ports stability adjusments. Take ballast in deep waters only (at least 200 mtrs of depth and 200 nm from nearest land). Avoid taking/discharge of ballast in ports or near shallow waters. 2. Carry-out annual ballast tank inspection.	Acceptable	1. Ballast exchange carried out on June/Oct - 2012 in Pacific Ocean. Next Ballast exchange to be planned/made on the First Quarter of the year 2013 prior arrival at Oceania port.	Chief Officer (Dec. 30, 2013)
9	Environmental Training	X	X	X	Increased understanding of Environmental protection.	Better knowledge & understanding	1. Computer Based Training(CBT) / Refresher courses ashore. 2. On board training - Waste/Garbage management.	Acceptable	1. Undergoing/ CBT #122 done by all crew. 2. Training carried out every Qtr - Mar, Jun, Sept, Dec'13.	Master (Dec. 30, 2013)
10	Oil Filters		X		Minimize Consumption.	No target set. Optimise consumption by sticking to the PMS and decreasing frequency as per the condition found but keeping safety in mind.	Proper monitoring as per planned maintenance system (PMS)	Acceptable	Cons. Jan-Dec.'12: 13 pcs	Chief Engineer (Dec. 30, 2013)

....contd

44

No.	Item			Objective	Target	Control Measures	Status	Records / Remarks	Responsible (Target Date)
11	Oily Rags		X	Minimize Consumption.	To reduce by 5 %.	1. By careful using of Rags in the machinery spaces. Place seperate drum for slighly soiled rags which can later be used for mopping up oil in scavange spaces etc.	Acceptable	Cons. 0.04 m3 per month. (same as last year)	Chief Engineer (Dec. 30, 2013)
12	Freon	x	X	Reduce pollution to environment	To reduce consumption by 5 %.	Gas leak detection to be carried out frequently so as to minimise chances of leakage. Machinery maintenance to be as per PMS.	Acceptable	R-417A Cons. Jan-Dec '12 : 56 Kg	Chief Engineer (Dec. 30, 2013)
13	Catering used chemicals, detergents, papers and plastics	X	X	X — Minimize Consumption of all catering Dept. used products.	To reduce consumption by 5 % based on the Set target for the past 12 months.	1.Close monitoring of consumption made. 2 .Proper usage on the amount of detergents, chemicals and other cleaning agents used.3. Proper disposal of all generated waste.	Acceptable	1. Records of consumption made for catering used products. 2. Target set for year 2012 is also indicated on the lower portion of the tally sheet made. 3. Detergent used in US waters are on Biodegradable type (kept in the store)	Catering Officer (Dec. 30, 2013)

45

Table 15. Objective with a Target of Reduction of the Volume of the Garbage Generated by 5% or More Compared to the Previous Year.

SNo	Garbage Categories	Total Volume Produced (Jan 2012 - Dec 2012) (m³)	Target Set (m³)	Target Met	Target Set for 2013, 5% Reduction	Responsibility and Measures
1	Hazardous waste, Plastic	32.5	35	Yes about 5% reduction	30 m³ 5% reduction	Master and crew
2	Floating dunnage, lining, or packing material	0.0	0.00	0.00	0.00	Continue hand over back lining and packing material to suppliers
3	Paper products, rags, glass, metal, bottle and crockery	1.96	2.05		1.86	Control unnecessary copying, using both sides
4	Food waste	3.33	*	*	*	*
5	Incinerated ash	0.05				
6	Used batteries	222 pc (0.122)	244	Yes	211	Use rechargeable
7	Aerosol	0.04				
8	Used florescent tubes and bulbs	725 pcs (1.33) 60.4 pcs/m	65 pc/m	Yes	58 pcs/m	Use good quality
9	Others like oily rags and oily waste	46.47		2		
GRAND TOTAL (m³)		85.81				

Table 16. Target to Reduce the Volume of Garbage Disposal to Shore Reception Facilities by 5 %, in Line with the Company's Environmental Management System Planning.

Volume of Garbage in M³			
Disposed to shore reception facilities	Jan-Dec 2012	Less 5%	Jan-Dec 2013
	75.1 M³	3.75 M³	71.35 M³

Figure 1. Solid and Hazardous Waste Management and Control System.

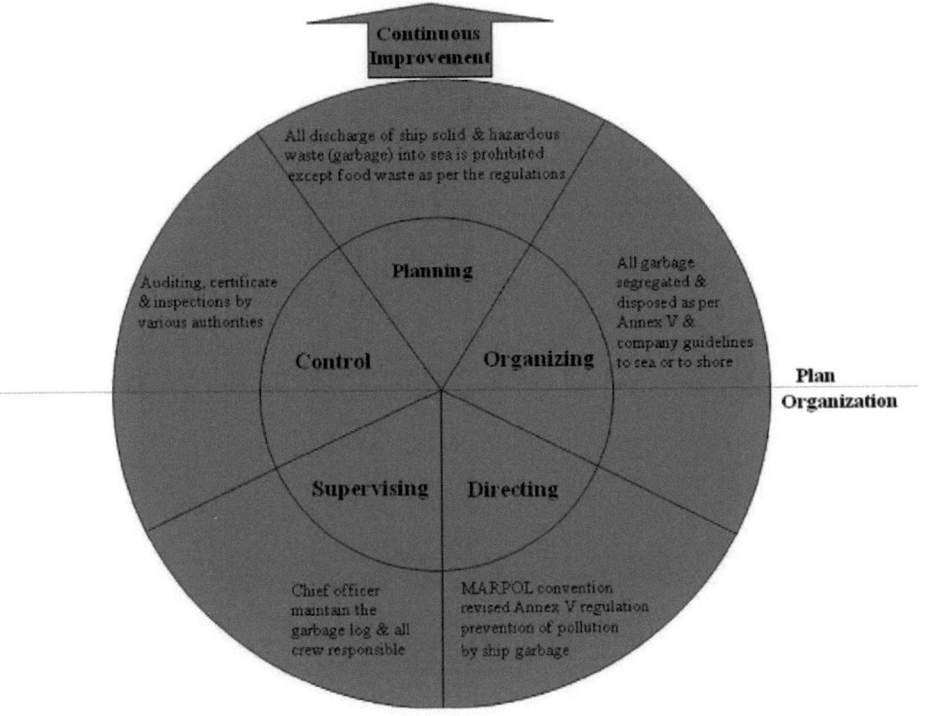

Figure 2. A Typical Organization Structure On board an Ocean Going Vessel.

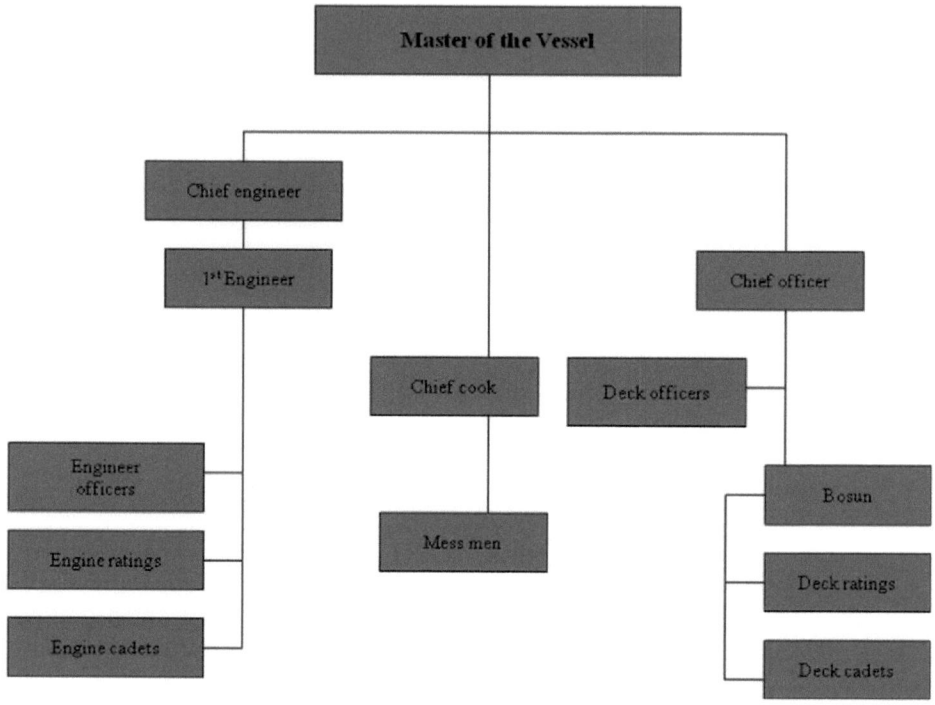

Figure 3. A Flow Chart for Ship Generated Solid and Hazardous Waste On Board a Merchant Vessel.

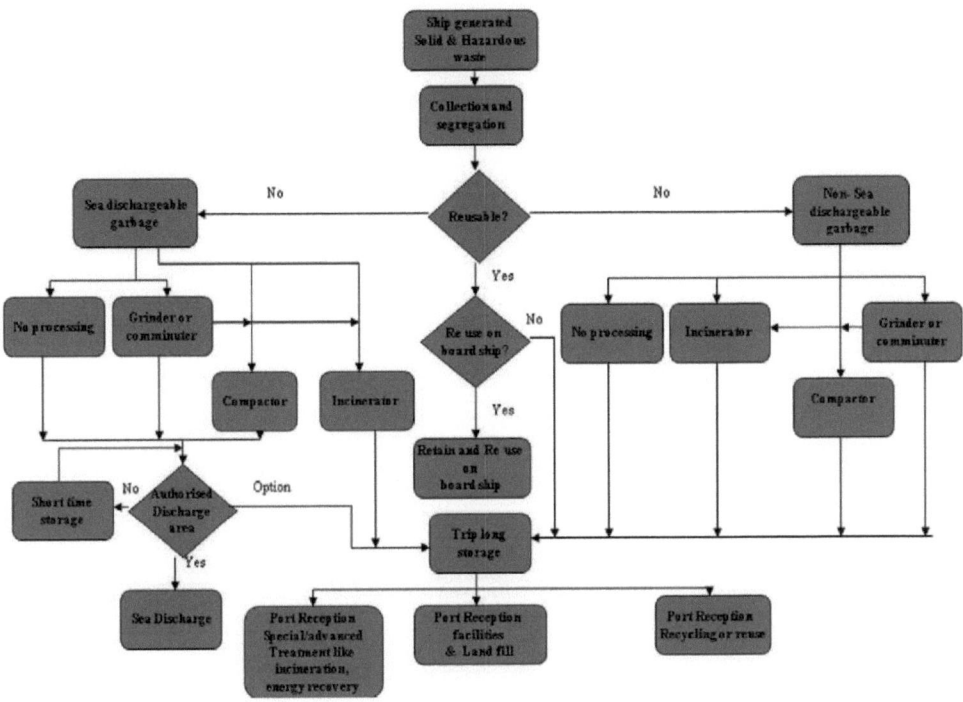

Figure 4. Solid and Hazardous Waste Management is Integral Part of Company's Environmental Management System.

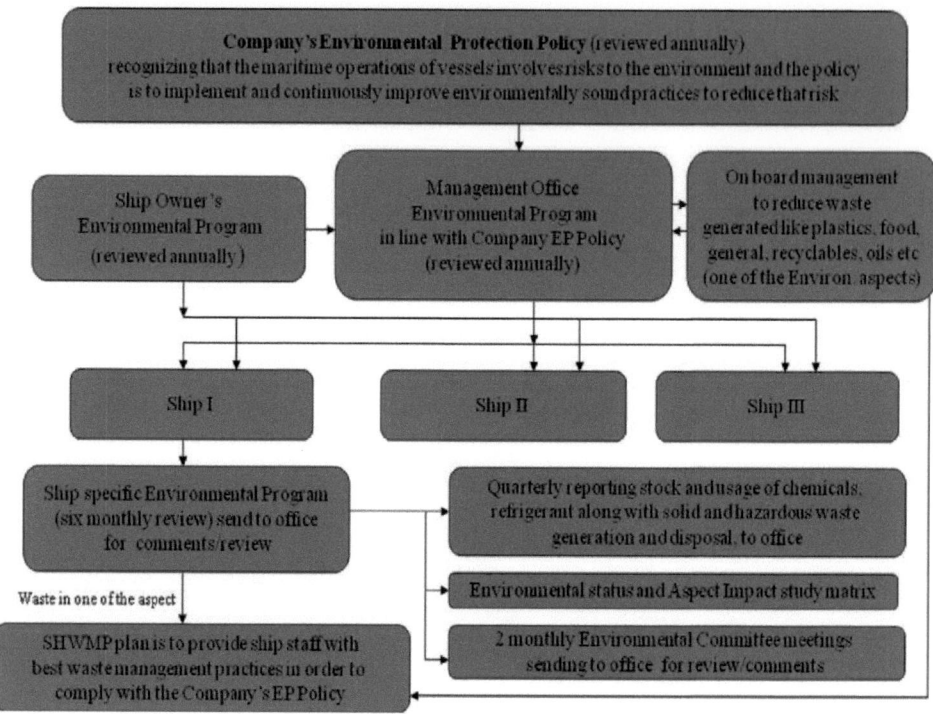

Appendix I. A Typical Format of Solid & Hazardous Waste (Garbage) Discharges Record: Before 01.01.2013.

RECORD OF GARBAGE DISCHARGES

Ship's name: MORNING CONCERT

Distinctive No., or letters: MNTZ5   IMO No.: 9312822

Garbage categories:

1: Plastics.

2: Floating dunnage, lining, or packing materials.

3: Ground paper products, rags, glass, metal, bottles, crockery, etc.

4: Cargo residues, paper products, rags, glass, metal, bottles, crockery, etc.

5: Food waste.

6: Incinerator ash except from plastic products which may contain toxic or heavy metal residues.

NOTE: THE DISCHARGE OF ANY GARBAGE OTHER THAN FOOD WASTE IS PROHIBITED IN SPECIAL AREAS. ONLY GARBAGE DISCHARGED INTO THE SEA MUST BE CATEGORIZED.
GARBAGE OTHER THAN CATEGORY 1 DISCHARGED TO RECEPTION FACILITIES NEED ONLY BE LISTED AS A TOTAL ESTIMATED AMOUNT.
DISCHARGES OF CARGO RESIDUES REQUIRE START AND STOP POSITIONS TO BE RECORDED.

Date /Time	Position of the ship	Estimated amount discharged into sea (m³)					Estimated amount discharged to reception facilities or to other ship (m³)	Estimated amount incinerated (m³)	Certification /Signature
		CAT.2	CAT.3	CAT.4	CAT.5	CAT.6	CAT.1 Other		
18 1800	N 16-50.8 W 069-08.2				0.010				⟋ %
19	ALL GARBAGE RETAINED ONBOARD								⟋ %
20	MANZANILLO INT'L TERMINAL	DISCHARGED GARBAGE TO SHORE FACILITY 2.3					0.7		⟋ %
21	PANAMA CANAL TRANSIT	ALL GARBAGE RETAINED ON BOARD							⟋ %
22 0915 0937	N 07-05.7 W 081-28.7 N 07-05.1 W 081-35.8							0.003	⟋ %
22 1800	N 08-18 W 083-47				0.020				⟋ %
23 0905 0927	N 10-20.9 W 087-51.8 N 10-23.8 W 087-57.7							0.005	⟋ %
23 1940	N 11-48 W 090-44				0.020				⟋ %
24 0939 1033	N 13-48.3 W 094.44.4 N 13-56 W 095-00.5							0.005	⟋ %
24 1848	N 15-03 W 097-15				0.015				⟋ %
25 1040 1119	N 17-10.6 N 101-22.3 N 17-15.6 W 101-44							0.004	⟋ %
26 0908 0945	N 20-37.2 W 107-44.7 N 20-43.3 W 107-55.2							0.007	⟋ %
26 1815	N 21-55.4 W 109-59.5				0.015				⟋ %
27 0853 1029	N 24-40.8 W 112-16.8 N 24-58.9 W 113-33.6							0.006	⟋ %
27 1900	N 26-57 W 115-08				0.012				⟋ %
28 1800	N 32-08.3 N 117-33				0.010				⟋ %
29	PORT HUENEME	— NO DISPOSAL/ALL GARBAGE RETAINED ONBOARD							⟋ %
30 0839 0943	N 86-36.6 W 123-05.5 N 37-16.8 W 123-16.9							0.008	⟋ %
31 0820 0905	N 43-24.1 W 125-38.6 N 43-24.5 W 125-38.9							0.003	⟋ %
31 1500	N 43-23 W 125-23				0.015				⟋ %
03 0810 0942	N 48-53.9 N 127-54 N 48-56.7 W 128-14.2							0.004	⟋ %

Master's signature: _Sanjay_   Date: 3rd June 2009

51

Appendix II. Typical Format of Solid & Hazardous Waste (Garbage) Discharges Record:
After 01.01.2013.

RECORD OF GARBAGE DISCHARGES

Ship's name: _____ TAKARA _____

Distinctive number or letters: __ LAZN 4 _____ IMO No.: __ 8506749 __

Garbage categories:

A. Plastics
B. Food wastes
C. Domestic wastes (e.g., paper products, rags, glass, metal, bottles, crockery, etc.)
D. Cooking oil
E. Incinerator Ashes
F. Operational wastes
G. Cargo residues
H. Animal Carcass(es)
I. Fishing gear

Date/ Time	Position of the ship/Remarks (e.g., accidental loss)	Category	Estimated amount discharged or incinerated (m3)	To sea (m3)	To reception facility (m3)	Incineration (m3)	Certification/ Signature
01-01-13 1900	32°28'N 163°22.7'E	B	0.02	0.02	–	–	Sav Uo
02-01-13 0900	32°41'N 167°48.3'S	C	0.10 (OILY RAGS)	–	–	0.10	M 2/6
02-01-13 1000	32°43'N 167°56.5'E		0.10 (PAPER)	–	–	0.10	M 2/6
02-01-13 1900	32°50.7'N 170°18.5'E	B	0.01	0.01	–	–	Sav Uo
03-01-13 1800	33°14.2'N 177°23.0'E	B	0.01	0.01	–	–	Sav Uo
03-01-13 1900	33°38.1'N 175°16.4'W	B	0.01	0.01	–	–	Sav Uo
04-01-13 1900	34°02.4'N 167°46.8'W	B	0.01	0.01	–	–	Sav Uo
05-01-13 1800	34°21.1'N 161°07'W	B	0.01	0.01	–	–	Sav Uo
06-01-13 1900	34°46.5'N 154°10.2'W	B	0.01	0.01	–	–	Sav Uo
07-01-13 1800	35°41.1'N 147°26.3'W	B	0.01	0.01	–	–	Sav Uo
08-01-13 1800	34°57.7'N 139°58.5'W	B	0.01	0.01	–	–	Sav Uo
09-01-13 1800	34°28.5'N 133°18.7'W	B	0.01	0.01	–	–	Sav Uo
10-01-13 1900	33°32.4'N 126°08.4'W	B	0.01	0.01	–	–	Sav Uo
11-01-13 1900	32°23.3'N 120°00.6'W	B	0.01	0.01	–	–	Sav Uo
14-01-13 1900	36°54'N 122°27 U WEST	B	0.03	0.03	–	–	Sav Uo
15-01-13 1800	41°24.1'N 128°19.8'W	B	0.01	0.01	–	–	Sav Uo
16-01-13 1900	46°24'N 134°57.9'W	B	0.01	0.01	–	–	Sav Uo
17-01-13 1900	50°29'N 142°17.5'W	B	0.01	0.01	–	–	Sav Uo
19-01-13 1800	54°N 151°13.14'W	B	0.01	0.01	–	–	Sav Uo

Master's signature: _____ Sign _____ Date: __ 19-01-2013 __

3

52

Appendix III. Advance Notification Form for Solid and Hazardous Waste Disposal to Shore Reception Facilities.

Notification of the Delivery of Waste to: _____
(enter name of port or terminal)

The master of a ship should forward the information below to the designated authority at least 24 hours in advance of arrival or upon departure of the previous port if the voyage is less than 24 hours.

This form should be retained on board the vessel along with the appropriate Oil Record Book, Cargo Record book or Garbage Record book.

DELIVERY FROM SHIPS (ANF)

1. SHIP PARTICULAR

1.1 Name of Ship:	1.5 Owner or Operator:
1.2 IMO Number:	1.6 Distinctive Number or Letter:
1.3 Gross Tonnage:	1.7 Flag State:
1.4 Type of Ship: ☐ Oil tanker ☐ Chemical tanker ☐ Bulk carrier ☐ Container ☐ Other cargo ship ☐ Passenger ship ☐ Ro-Ro ☒ Other (Car Carrier)	

2. PORT AND VOYAGE PARTICULARS

2.1 Location / Terminal Name & POC:	2.6 Last Port where waste was delivered:
2.2 Arrival Date & Time:	2.7 Date of Last delivery:
2.3 Departure Date and Time:	2.8 Next Port of delivery (if known):
2.4 Last Port & Country:	2.9 Person submitting this form is
2.5 Next Port & Country (if known):	(if other than the master):

3. TYPE AND AMOUNT OF WASTE FOR DISCHARGE TO FACILITY

MARPOL Annex I – Oil	Quantity (m³)	MARPOL Annex V - Garbage	Quantity (m3)
Oily bilge water		Plastic	
Oily residues (sludge)		Floating dunnage, lining or packing materials	
Oily tank washing		Ground down paper products, rags, glass, metal, bottles, crockery etc	
Dirty ballast water		Cargo residues**, paper products, rags, glass, metal, bottles, crockery, etc.	
Scale and sludge from tank cleaning		Food waste	
Other (please specify)		Incinerator ash	
MARPOL Annex II - NLS	**Quantity (m³)/ Name ***	Other waste (specify)	
Category X substance		**MARPOL Annex VI - Air pollution**	**Quantity (m³)**
Category Y substance		Ozone-depleting substances and equipment containing	

			such substance	
Category Z substance			Exhaust gas – cleaning residues	
OS – other substance				
MARPOL Annex IV - Sewage	**Quantity (m³)**			
Name of Ship:			IMO Number:	

Please state below the approximate amount of waste and residues remaining on board and the percentage of maximum storage capacity. If delivering all waste on board at this port please strike through this table and tick the box below. If delivering some or no waste, please complete all columns.

I confirm that I am delivering all the waste held on board this vessel (as shown on page 1) at this port ☐

Type	Max dedicated stora Capacity m3	Amount of waste retained on board m	Port at which remaining waste wil be delivered (if known)	Estimate amount of waste to be generate between notification and next port of call m3
MARPOL Annex I - Oil				
Oily bilge water				
Oily residues (sludge)				
Oily tank washing				
Dirty ballast water				
Scale and sludge from tank cleaning				
Other (please specify)				
MARPOL Annex II - NLS *				
Category X substance				
Category Y substance				
Category Z substance				
OS – other substance				
MARPOL Annex IV - Sewage				
Sewage				
MARPOL Annex V - Garbage				
Plastic				
Floating dunnage, lining or packing materials				

Ground paper products, rags, glass, metal, bottles, crockery				
Cargo residues**, paper products, rags, glass, metal bottles, crockery.				
Food waste				
Incinerator ash				
Other waste (specify)				

Date: _____   **Name & Position:** _____

Time : _____   **Signature :** _____

Appendix IV. Notification Form for Non-Availability of Sufficient /Proper Shore Reception Facilities for Solid and Hazardous Waste Disposal.

The master is obliged under MARPOL Annex V requirements to report any non-availability of reception facilities at any port using the IMO standard format below.

The Master of a ship having encountered difficulties in discharging waste to reception facilities should forward the information below, together with any supporting documentation, to the Administration of the flag State and, if possible, to the competent Authorities in the port State.

1   SHIPS PARTICULARS

1.1 Name of Ship: _____

1.2 Owner or Operator: _____

1.3 Distinctive number or Letters: _____

1.4 IMO Number: _____

1.5 Gross Tonnage: _____

1.6 Port of Registry: _____

1.7 Flag State: _____

1.8 Type of Ship: _____

Oil tanker Chemical tanker Bulk carrier Other cargo ship Passenger ship Other (specify)

2   PORT PARTICULARS

2.1 Country: _____

2.2 Name of Port or Area: _____

2.3 Location/Terminal name: _____ (e.g. berth/terminal/jetty)

2.4 Name of company operating the reception facility (if applicable):

_____

2.5 Type of Port Operation: _____

Unloading port Loading port Shipyard Other (specify)

2.6 Date of arrival: __/__/____ (dd/mm/yyyy)

2.7 Date of occurrence: __/__/____ (dd/mm/yyyy)

2.8 Date of departure: __/__/____ (dd/mm/yyyy)

3   INADEQUACIES OF FACILITIES

3.1 Type and amount of waste for which the port reception facility was inadequate and nature of problems encountered

Type of Waste	Amount for Discharge (m³)	Amount not Accepted (m³)	Problems Encountered Indicate the problems encountered by using one or more of the following code letters, as appropriate. A. No facility available B. Undue delay C. Use of facility technically not possible D. Inconvenient location E. Vessel had to shift berth involving delay/cost F. Unreasonable charges for use of facilities G. Other (please specify in paragraph 3.2)
MARPOL Annex I-related			
Type of oily waste:			
Oily bilge water			
Oily residues (sludge)			
Oily tank washings (slops)			
Dirty ballast water			
Scale and sludge from tank cleaning			
Other (please specify .....)			
MARPOL Annex II-related			
Category of NLS$_1$ residue/water mixture for discharge to facility from tank washings:			
Category X substance			
Category Y substance			
Category Z substance			
MARPOL Annex IV-related Sewage			
MARPOL Annex V-related Type of garbage:			
Plastic			
Floating dunnage, lining, or packing materials			
Ground paper products, rags, glass, metal, bottles, crockery, etc.			
Cargo residues, paper products, rags, glass, metal, bottles, crockery, etc.			
Food waste			
Incinerator, ash			
Other (please specify .....)			
MARPOL Annex VI-related			
Ozone depleting substances and equipment containing such substances			
Exhaust gas cleaning residues			

3.2 Additional information with regard to the problems identified in the above table.

_____

_____

_____

_____

_____

3.3 Did you discuss these problems or report them to the port reception facility?

Yes   No

If *Yes*, with whom (please specify)

_____

_____

If *Yes*, what was the response of the port reception facility to your concerns?

_____

_____

3.4 Did you give prior notification (in accordance with relevant port requirements) about the vessel's requirements for reception facilities? *Yes   No   Not applicable*

If *yes*, did you receive confirmation on the availability of reception facilities on arrival? *Yes   No*

4   ADDITIONAL REMARKS/COMMENTS

_____

_____

_____

_____

_____

Master's Signature   **Date: __/__/____ (dd/mm/yyyy)**